Author Index

69th Conference
on Glass Problems

69th Conference on Glass Problems

A Collection of Papers Presented at the
69th Conference on Glass Problems
The Ohio State University, Columbus, Ohio
November 4–5, 2008

Edited by
Charles H. Drummond, III

A John Wiley & Sons, Inc., Publication

Published by John Wiley & Sons, Inc., Hoboken, New Jersey.
Published simultaneously in Canada.

For general information on our other products and services or for technical support, please contact our Customer Care Department within the United States at (800) 762-2974, outside the United States at (317) 572-3993 or fax (317) 572-4002.

Wiley also publishes its books in a variety of electronic formats. Some content that appears in print may not be available in electronic format. For information about Wiley products, visit our web site at www.wiley.com.

Library of Congress Cataloging-in-Publication Data is available.

ISBN 978-0-470-45751-1
ISBN 978-0-470-52427-5 (special edition)

Printed in the United States of America.

10 9 8 7 6 5 4 3 2 1

Contents

COMBUSTION AND ENERGY SAVINGS

ENVIRONMENTAL ISSUES AND NEW PRODUCTS

Foreword

The 69th Conference on Glass Problems was sponsored by the Departments of Materials Science and Engineering at The Ohio State University. The director of the conference was Charles H. Drummond, III, Associate Professor, Department of Materials Science and Engineering, The Ohio State University.

Gregory N. Washington, Interim Dean, College of Engineering, The Ohio State University, gave the welcoming address. Rudolph Buchheit, Chair, Department of Materials Science and Engineering, The Ohio State University, presented the Departmental welcome.

The themes and chairs of the five sessions were as follows:

MELTING AND MODELING
Terry Berg, CertainTeed, Athens GA, Martin H. Goller, Corning, Corning NY and Phil Ross, Glass Industry Consulting, Laguna Niguel CA

REFRACTORIES
Dick Bennett, Johns Manville, Littleton CO, John Tracey, North American Refractories, Cincinnati OH and Carsten Weinhold, Schott, Duryea PA

COMBUSTION AND ENERGY SAVINGS
Sho Kobayashi, Praxair, Danbury CT, Larry McCloskey, Toledo Engineering, Toledo OH and Elmer Sperry, Libbey Glass, Toledo OH

ENVIRONMENTAL ISSUES AND NEW PRODUCTS
Ruud Beerkens, TNO Glass Technology - Glass Group, Eindhoven, The Netherlands, Tom Dankert, Owens-Illinois, Toledo OH and Gerald DiGiampaolo, PPG Industries, Pittsburgh PA

Preface

In the tradition of previous conferences, started in 1934 at the University of Illinois, the papers presented at the 69th Annual Conference on Glass Problems have been collected and published as the 2008 edition of The Collected Papers.

The manuscripts are reproduced as furnished by the authors, but were reviewed prior to presentation by the respective session chairs. Their assistance is greatly appreciated. C. H. Drummond did minor editing with further formatting by The American Ceramic Society. The Ohio State University is not responsible for the statements and opinions expressed in this publication.

CHARLES H. DRUMMOND, III
Columbus, OH

Acknowledgments

It is a pleasure to acknowledge the assistance and advice provided by the members of Program Advisory Committee in reviewing the presentations and the planning of the program:

Ruud G. C. Beerkens—TNO

Dick Bennett— Johns Manville

Terry Berg—CertainTeed

Tom Dankert—Owens-Illinois

Gerald DiGiampaolo—PPG Industries

Martin H. Goller—Corning

H. "Sho" Kobayashi—Praxair

Larry McCloskey—Toledo Engineering

C. Philip Ross—Glass Industry Consulting

Elmer Sperry—Libbey Glass

John Tracey—North American Refractories

Carsten Weinhold—Schott

Melting and Modeling

EXPERIENCES WITH AN OXYGEN-FIRED CONTAINER GLASS FURNACE WITH SILICA CROWN – 14 YEARS – A WORLD RECORD?

J.J. Schep
O-I The Netherlands

ABSTRACT
Fourteen years ago the introduction of oxy-fuel combustion in glass melting technology in the Dutch glass industry was started. This paper refers to the first oxy-fuel furnace in The Netherlands, and the experiences with a silica crown. First an explanation about the environmental legislation in The Netherlands is given, which lead to the installation of this first oxy-fuel furnace. This is followed by a description of this first full scale oxy-fuel implementation and the experiences with emission reduction, energy consumption, refractory wear and glass composition in this furnace.

DUTCH ENVIRONMENTAL REGULATIONS:
At the end of the 1980's the government in the Netherlands made, compared to European standards, extensive plans to protect the environment and stimulating a more efficient use of our natural resources. The national environmental plans focused on the reduction of the emissions of components that cause acid rain like SO_x and NO_x, but also dust emissions and energy consumption should be decreased. Based on these plans the Netherlands Emission Regulations (NeR), with a special glass paragraph were developed in 1992-1993 and also the glass industry was invited to make a multiple year agreement to improve in a systematically and programmed way the energy efficiency of their processes.

In the early 90's the Dutch glass industry was already faced with large investments for end of pipe technologies in case NOx-emissions had to be reduced to levels which can not be reached by conventional primary measures only. At that same time, increasingly, oxy-fuel technology was developed and implemented in the United States. The Dutch glass industry, united in the VNG, decided to investigate, in cooperation with TNO-TPD in Eindhoven (Prof. Beerkens), the possibilities of introducing oxy-fuel technology in the Netherlands. A trip was made to the United States to visit several glass companies that already implemented oxy-fuel combustion technology.

The glass industry discovered in oxy-fuel a process integrated technology for the reduction of NOx, which could also improve the energy performance of the furnace and/or glass quality and melting loads could be increased by application of oxygen firing. This in contrast with the end of pipe technologies, like SCR and SNCR, which did not add to the performance and efficiency of the furnace and therefore could never be cost effective.

Because of the process-integrated character, the government was also interested in the oxy-fuel technology. They were willing to finance a demonstration project with a full-scale industrial oxy-fuel fired glass-melting furnace. The conditions of this financial support were that experiences with oxy-fuel should be shared with the other companies within the Dutch glass industry.

The glass industry was given the freedom, within the NeR, to choose between two different tracks to reduce the emission levels: First the reduction of NO_x-emission by implementing oxy-fuel (NO_x down to less than 1 kg/ton of molten glass), a technique that was developing in those days for most glass industries, followed by the reduction particulate emissions by the introduction of dust filters or electrostatic precipitators. The other track contains the same measures only in a different order. Mentioned in

these regulations was, that instead of oxy-fuel firing improved lonox measures also could be implemented, if this would lead to similar emission values.

Both tracks should be completed in 2010, with an intermediate deadline in 2003 for either one of the measures. Negotiated was the fact that these extensive measurements could be implemented only at major cold repairs of the glass melting furnaces.

THE FIRST OXY-FUEL FURNACE:
In 1994 the L4 container glass furnace of O-I, at that time N.V. Vereenigde Glasfabrieken, was up for a rebuilt. The existing recuperative furnace was old and had a poor energy-efficiency. Also the building was not suitable to hold a regenerative furnace, either end or side-port, for the required production volume. For these reasons and the stricter emission limits it was decided to build an oxy-fuel furnace. For this ambitious project several partners were sought.

For the furnace design, Owens Brockway was consulted, because of their long-term experience in developing new furnaces. They supplied the design of the furnace and the exhaust system. Owens-Brockway engineered the furnace steel and refractory and made the specifications of the flue gas system.

The flue system is designed to cool down the flue gases gradually to avoid early condensation of sulphates and protecting the stack, because there was no heat recovery system installed like regenerators. They also made a heat balance with their recently developed heat balance program for oxy-fuel furnaces.

The expertise for the on site production of oxygen and the supply of the oxygen (valve trains, burners, etc.) was delivered by Air Liquide. They supplied a VSA-unit (Vacuum Swing Absorption-unit) with a capacity of 2100 nm^3/h oxygen (calculated to 100%) with a purity of 91 % ± 2%. Later this capacity was reduced to 1500 nm3/h, to use the installation more efficiently. To ensure uninterrupted oxygen supply a liquid oxygen tank with a capacity of 70.000 litres and evaporators was installed.

Figure 1: VSA-unit

Figure 2: Back-up tank and evapora-

They also supplied the burners (1st generation pipe in pipe) and burner control system. In addition the oxygen supplier provided a model calculation of the combustion space with their ATHENA ® modelling tool. See figure 3 for an example of their modelling of the crown temperatures and the actual temperatures.

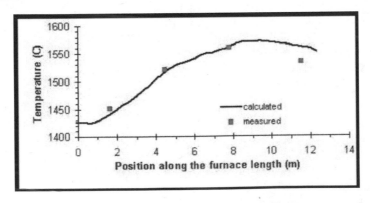

Figure 3: Calculation crown temperatures.

The furnace:
During the rebuilt of the L4 the melting area of the furnace was increased with 17% while the melting capacity increased with more than 25%. (see table 1)

Table 1: Furnace conversion data.

	Unit melter	Oxy-fuel
Melter Area	64 m^2	75 m^2
Length : Width ratio	1,9 :1	2 :1
Glass depth	0,9 m	0,9 m
Fuel	Natural gas	Natural gas
Boosting capacity	800 kW	800 kW
Burners	8 per side (opposite)	6 per side (staggered)
Burner capacity		6 of 1MW 6 of 0,5 MW
Melting capacity	136 ton/day	>200 ton/day
Glass colours	Amber, emerald green	Amber, emerald green
Products	Beer bottles, jars	Beer bottles, jars

The furnace was build in April 1994 and has been in production since. This makes it probably the longest running all oxy-fuel fired furnace, without a cold repair, in the world.

ENERGY EFFICIENCY:
The energy performance of the old L4 was poor (7200 MJ/ton), due to the low air preheat temperatures and poor insulation. The efficiency was improved with about 50%. To get a fair comparison, the specific energy demand was compared with 2 end-port furnaces (table 2). The end-ports used in the comparison belong to the most energy efficient furnaces in the world, which was shown by a benchmarking study in the Netherlands, done by TNO Eindhoven, which included over 100 furnaces world wide. The results are shown in figure 4.

Table 2. Furnace comparison data.

	Oxy fuel	End-port	End-port
E-Boost	YES	YES	NO
Glass colour	amber	green	amber/green
% cullet	77 %	79 %	74%
Pull (T/m2/day)	2,9	2,9	2,6

Compared to a modern end-port furnace, you can see that although energy is needed for the production of oxygen, the efficiency of an oxy-fuel furnace is comparable to a conventional end-port furnace. In this case the electrical power is not is not recalculated to primary energy.

However, since the costs of electrical power in The Netherlands are higher than those for Natural gas you can see in the same graph that the oxy-fuel is less cost efficient.

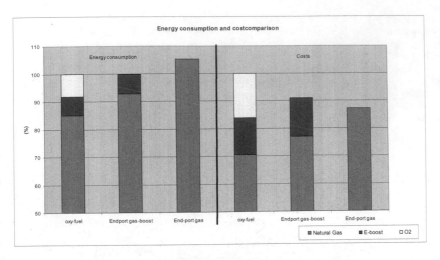

Figure 4: Energy consumption and energy cost

Over the years several comparisons were made to investigate the advantages and disadvantages of oxy-fuel firing. The ageing of the furnace was compared to the end-port furnaces, which were used for the energy comparison. As one can see in the graph below the ageing of an oxy-fuel furnace is less than that of a conventional furnace. As conventional end-port show an energy consumption increase of about 1 to 1,5% per year, the oxy-fuel furnace showed no ageing. The main reason is the lack of regenerators which loose efficiency because of fouling.

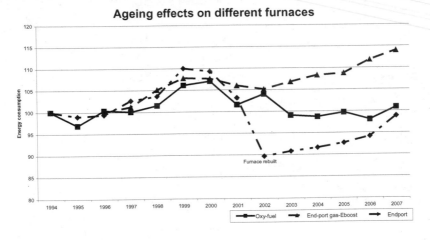

Figure 5: Ageing comparison between an oxy-fuel furnace and 2 end-

ENVIRONMENTAL ISSUES:
Comparing the emission data (see table 3), you can see much lower NO_x-emission levels for the oxy-fuel furnace compared to the "air" fired furnace. The higher SO_x-emission and dust emission is mainly caused by the colour of glass produced in the oxy-fuel furnace with higher surface temperatures and high sulphate in the batch. Also the higher flame temperature and water content in the flue gases cause increased evaporation rates and concentrations of glass components in the flue gases are higher. The figures in this table are given in kg/ton of molten glass, because of the lower flue gas volume. Because of the much smaller volumes of flue gas (hardly any nitrogen from air) per ton molten glass, the concentrations of the emitted components are higher than the flue gas concentrations in air-fuel fired furnaces. Typically the amount of flue gas per ton molten glass decreases by a factor 3-4.

Table 3: Environmental data

Component	Oxy-fuel	End-port
NO_x	< 0,90	< 1,80
SO_x	< 1,10*	< 0,60
Dust	< 0,20	< 0,10 (without filter)
F	< 0,004	< 0,001
CL	< 0,05	< 0,013
Lead	< 0,01	< 0,01

* Due to higher input of sulphate required to obtain the desired glass colour

The NO_x-emissions were influenced over the years by increased infiltration of cold air, either coming from the cooling wind or the furnace camera. Proper sealing of the joint between the sidewalls and the tuck stones and conscious use of camera purging air reduced the influences of these factors.

Because Dutch (low calorific) natural gas contains 12-14 % of nitrogen, lower values than the furnace emitted were at that moment difficult to reach. With high calorific gas (no N_2) and 100 % oxygen it is possible to reach NOx-emission values lower than 0,6 kg/ton of glass.

Recent developments in the conventional furnaces (lonox burners, port design) make it possible to reach similar emission levels for NO_x as oxy-fuel firing as well as similar energy consumption.

REFRACTORY WEAR:
A short time after start-up the first problems with superstructure refractory wear occurred. Dripping of refractory corrosion products was observed in front of the burners and peepholes. Also grooves originating from the silica crown became visible. After a few months wear on joints in the silica became visible.

All this resulted in a stoning problem in the finished product. Due to the chemical composition of the stones, they were called "Zacoline." Investigations were started and the flue gas composition was checked. Due to the lower flue gas volume, the concentration of volatile components in the flue gas was found to be 3-5 times higher compared to a conventional furnace, especially sodium hydroxide. Also the water content of the flue gasses was higher, 50% compared to 18% for a conventional furnace, because of the absence of large quantities of nitrogen in the combustion process.

Also the required crown temperatures for good melting appeared to be very low (<1400 °C) at that time, because the heat transfer to the glass bath was excellent. All this resulted in a deep penetration and reaction of sodium hydroxide with the silica from the silica crown, forming low temperature melt-

ing sodium silicates. The reaction between sodium hydroxide vapours originating from the melt and silica takes place below a certain critical temperature, depending on the sodium hydroxide concentration. In oxygen-fired furnaces the NaOH concentration is very high and this limits the temperature window for the crown: the temperature has to be controlled between 1460 and 1600 °C. In an air fired furnace the window is wider, typically: 1380-1600 °C.

In the colder areas of the silica crown the silica is attacked by NaOH vapours forming rat holes and relatively cold open joints, this will enhance the reaction process. The reaction products can melt during temperature changes.

During pull changes the temperature of the crown changed and silica melted of the crown and was running over the AZS material of the superstructure. This resulted in the grooves in the AZS and stones containing Al_2O_3, SiO_2 and ZrO_2.

After these observations and experiences the crown temperature was kept constant above 1470 °C. Also the temperatures were kept stable during pull changes.
This resulted in a decrease of silica wear, but it never stopped completely, which resulted in extensive maintenance over the years. Since 1997, every year, some hot repair work has been done on the crown, mainly ceramic welding and a complete overcoat of the crown with a concrete layer. (figures 6 and 7)

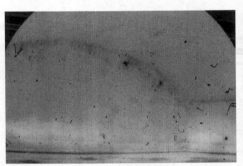 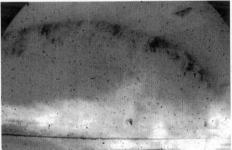

Figure 6: Condition crown and bridgewall in 2002.   Figure 6: Condition crown and bridgewall in 2002.

OPERATING THE FURNACE:
Because oxy-fuel firing provided a very new technology to the glass industry in the Netherlands, people in the plant had to learn a new set of safety rules, coming from working with oxygen. Extra attention was needed, to work with clean tools and clothing, to make sure no dust or oil residue would come into contact with the oxygen.

The furnace was also equipped with extra safety features to make sure that firing the furnace with oxygen was safe. These safety features included extra control of refractory temperature (extra thermocouples in breast walls and flue gas channel) and furnace pressure. Alarms would trigger the combustion system to run in so called "idle" mode (reduced firing with less burners) or to a complete shutdown.

Because external safety is a major concern of all companies, extra safety measures had to be taken for the storage of the liquid oxygen. All oxygen installation are checked on a yearly basis by an outside

company specialized in this kind of inspections. Thanks to all the measures taken and responsible behaviour, no issues with safety have been report in the last 14 years.

Looking at the costs, it is obvious that making oxygen compared to the use of air as an oxidant has its price. Calculations for this furnace were made to investigate the economic feasibility of an oxy-fuel furnace compared to a conventional furnace. The oxy-fuel option was, in this case, compared to a conventional end-port furnace, taking into account the reduced investment (no regenerators, no construction changes in the building), reduced construction time and the energy savings, but also the variable costs for the oxygen production. At that time with the funding of the government, the project was feasible.

At this moment we can see, from our practice that by proper maintenance, good control and early repairs of open joints or holes, it is possible to have a similar lifetime like conventional furnaces of over12 years.

GLASS PROPERTIES
Beside the stoning problems, which were mentioned before, there were some glass property related issues. The water content in the glass changed to a higher level after the conversion of the recuperative furnace to an oxygen fired one. This was caused by the higher water vapour pressure above the melt. The higher water content in the melt had some influence on the workability of the glass. The viscosity of the glass was not the same and changes in the batch composition were necessary to correct this. Some batch modifications appeared also to be necessary in order to stabilize the glass colour in the oxygen fired furnace. Due to different redox conditions in the glass and combustion space, salt cake and carbon amounts and ratios had to be changed.

OTHER OXY-FUEL EXPERIENCES AT O-I IN THE NETHERLANDS:
In 1997, O-I build a second oxy-fuel furnace in its Schiedam plant, similar to the 1st oxy-fuel furnace, with some improvements in the refractory design and choice of materials, except for the crown material. Burners, controls and oxygen supply are the same as in the first oxy fuel furnace.

Experiences in the second oxy-fuel furnace of O-I show that design changes and a different material choice as well as steady control (from the start) of the furnace temperatures can reduce the refractory wear. Still small joints can cause refractory problems and holes should be repaired as quickly as possible.
After 11 years the first extensive crown repair has been done. Damage to the crown was caused by fluctuating conditions as a result of a plant fire and power failure.

CONCLUSIONS:
Looking at the experiences in the past 14 years, oxy-fuel could be considered as a possible alternative for the reduction of NO_x-emissions for the future when the economical issues are resolved.

Although not all the refractory problems are completely solved we can conclude that oxy-fuel is an acceptable alternative to reduce NO_x-emissions below 1 kg/ton of molten glass in the container glass sector. Even with a silica crown acceptable life times can be reached when process conditions are kept stable and the furnace is maintained timely and properly.

The refractory problems are limited mainly to the combustion space and this can be reduced by improved design, steady conditions and proper maintenance.

An oxy-fuel furnace can be run for a longer period of time, even with a silica crown.

A combination of the right design and handling of the furnace can lead to lifetimes longer than conventional furnaces, because of the steady conditions wear on other parts like sidewalls can also be reduced.

Observations in the L4 showed that glass problems can be solved by adaptations of the batch and steady conditions in the furnace.

Still one issue keeps unresolved to make oxy-fuel furnaces the successful solution for the future. These are the costs for the production of oxygen. For this, a full conversion to oxy-fuel should be checked on economic feasibility in each individual case.

FINING OF GLASS MELTS: WHAT WE KNOW ABOUT FINING PROCESSES TODAY

Ruud Beerkens
TNO Science & Industry
The Netherlands

ABSTRACT

The paper addresses the mechanisms of fining (removal of gases from melt) and the effect of batch composition, oxidation state of the melt and furnace atmosphere on bubble removal processes for commercial glass types, such as float glass and container glass compositions. The mechanisms of the different stages of sulfate chemistry in the batch and sulfate fining process are presented, depending on the level of addition of cokes in the batch. The fining gas release as function of temperature from batch and melt are shown for different soda-lime-silica glass forming raw material batches. In case of coke and sulfate containing batches, sulfur dioxide gas release takes place in steps and at different temperature levels. SO_2 may evolve from the batch blanket (at about 900 $^{\circ}$C), but also at higher temperatures from the fresh melt starting at about 1050 $^{\circ}$C up to 1250 $^{\circ}$C, some SO_2 release can take place up to 1350 $^{\circ}$C and then for oxidized melts a strong evolution between 1420 and 1500 $^{\circ}$C. The last stage of SO_2 evolution from the melt is generally accompanied with oxygen gas formation as well. Bubble growth rates during primary fining a float glass melts from experiments and models are compared and the removal of dissolved gases from such melts will be shown.

The paper shows that furnace atmosphere may have an important influence on the temperature of fining gas evolution and total fining gas production: water vapor or helium in the atmosphere will reduce fining onset temperature and may significantly increase gas evolution during primary fining.

1. INTRODUCTION

The removal of gases from glass melts is not limited to the elimination of bubbles, blisters and seeds from the molten glass, but includes also stripping of dissolved gases from glass melts. Effective stripping of gases such as nitrogen and CO_2, will reduce the risk of reboil and blister formation downstream the primary fining process in a glass melt tank, for instance by interaction of the melt with refractory materials (refractory surfaces may act as nucleation sites for new bubbles in saturated or super-saturated melts).

The mechanism of primary fining, which comprises essentially the removal of bubbles by bubble growth and enhanced bubble ascension in the melt in combination with gas stripping and the re-absorption of bubbles during controlled cooling of the melt (secondary fining), will be presented and discussed. The behavior of gas bubbles in laboratory melts in transparent vitreous silica crucibles during both stages can be observed by video movies and gas evolution or bubble growth can be studied for different batch compositions, heating rates or furnace atmospheres[1].

After batch melting, typically 10^5 to 10^6 bubbles (diameter 0.05-0.4 mm) and seeds per kg molten glass may be formed in industrial furnaces[2] and all these bubbles have to be removed in high quality glass production. It is estimated that the volume of bubbles, just of batch melting-in is only about 1 % of the glass melt volume (estimated from Mulfinger et al[2]).

The controlled release of fining gases, essential for the primary fining process depends on temperature, fining agent content of the batch & melt and oxidation state of batch & melt. The fining gas evolution by the aid of sulfate and carbon will be shown for different redox states of the melt and batch. Sulfate – carbon reactions in the batch blanket will also influence the stages of gas evolution from the obtained melt. The application of evolved gas analysis during heating and fining of the glass melts will give information on the fining stages and activity of the fining agent as a function of temperature.

Furthermore, the furnace atmosphere above the melt will influence the bubble growth rates as both laboratory experiments and modeling studies show. Especially helium and water vapor will enhance fining of glass melts but may also lead to increased foaming. Often adjustments to the batch are necessary, when changing to other furnace atmospheres, for instance after conversion from air to oxygen firing.

In this paper, the sulfate-carbon batch reaction mechanisms, the evolved gases during heating, melting and fining of sulfate containing soda-lime-silica glass melts are shown and discussed. The bubble growth mechanism and bubble growth modeling and experiments are presented and the impact of the furnace atmosphere on fining and bubble growth will be explained.

2. BATCH REACTIONS OF SULFATES

Sulfates are added to many soda-lime-silica batches (* including other raw materials such as dolomite, iron oxides, iron blast furnace slags, feldspars, coloring agents and cokes) to aid the fining process by evolution of fining gases such as SO_2 at elevated melt temperatures and low glass melt viscosity levels. Sulfates also will reduce the surface tension of the primary melting phases in the heated batch blankets in glass furnaces. The lowering of the surface tension of these primary melts will increase the surface area of the sand grains that are in contact with these reacting alkali-rich phases and thus enhances reactive sand grain dissolution[3].

During batch melting, the batch forms its own internal atmosphere, resulting from batch gases that are released from the batch during the reactions upon heating. These gases include water vapor, CO_2, CO and in case of sulfates in the batch: SO_2 plus oxygen or NO/NO_2, in case of nitrates. Many batch reactions are associated with the formation of gas (e.g. carbonate decomposition or reactions of soda with sand forming CO_2 gas). Gas evolution during heating and during reactive melting (** batch components react and form reaction products that may be liquid) of the batch can be characteristic for the raw material-melt conversion steps and the occurring sulfate reactions. This gas evolution can be monitored during laboratory batch melting experiments, using gas chromatography, FTIR analysis (Fourier Transform Infra-Red analysis) or mass spectrometric analysis of gases exiting from the melting tests. This method is called Evolved Gas Analysis. EGA for different gas species can be measured: CO, CO_2, O_2, SO_2, etcetera. Generally, the release of a type of gas species as a function of time or temperature is measured and temperature intervals, corresponding with largest evolution rates of certain gas species, refer to batch-reactions that are accompanied with these released gases. For soda-lime-silica batches, used for container or float glass production, at temperatures above 1100 °C, most batch is transferred into molten glass. Gases released above this temperature level are caused by gas formation during decomposition (sulfates, antimony oxides) or evaporation (NaCl) of fining agents dissolved in the melt.

Addition of organic components to the batch will lead to coke formation (pyrolysis of organic components will form char) during heating of the batch as soon as the batch is covered by phases hindering the entrance of oxygen into the batch.

Coke can react directly with oxidizing components in the batch such as nitrates or sulfates or ferric iron (Fe^{3+} or Fe_2O_3). In case of sodium sulfate particles / powder in contact with cokes the following reactions may take place:

$$C \text{ (batch)} + Na_2SO_4 \text{ (batch)} \rightarrow CO \text{ (gas)} + Na_2O \text{ (batch)} + SO_2 \text{(gas)} \qquad (1a)$$
$$C \text{ (batch)} + 2Na_2SO_4 \text{ (batch)} \rightarrow CO_2 \text{ (gas)} + 2Na_2O \text{ (batch)} + 2SO_2 \text{(gas)} \qquad (1b)$$

Or at very high carbon contents sulfide may be formed:

$$4C \text{ (batch)} + Na_2SO_4 \text{ (batch)} \rightarrow 4CO \text{ (gas)} + Na_2S \text{ (batch)} \qquad (2a)$$

$4CO$ (batch) $+ Na_2SO_4$ (batch) $\rightarrow$ $4CO_2$ (gas) $+ Na_2S$ (batch) (2b)

Tammann and Oelsen[4] already found evidence for formation of sulfide in the batch containing cokes. The components Na_2S and Na_2SO_4 may form an eutectic melt between 740 and 800 °C. This melt is rather reactive and promotes batch melting.
Carbon may also reduce iron oxides in the batch:

Fe_2O_3 (batch) $+ C$ (batch) $\rightarrow 2FeO$ (batch) $+ CO$ (gas) (3)

These reactions 1 to 3 are solid state reactions and the overall conversion is limited due to the limited contacts between carbon particles and sulfates or iron oxides.

Evolved gas analysis however shows that most CO evolution takes place at the same time as decomposition of the carbonates occur. According to references[5, 6], this is due to the so-called Boudouard reaction taking place in the gas phases inside the batch blanket during the CO_2 evolution from the carbonates (lime, dolomite, soda). The large amounts of released CO_2 (up to 200 grams per 1 kg glass) during batch heating and melting, react with the carbon/cokes particles in the batch:

CO_2(gas) $+ C$ (batch) $\Leftrightarrow 2CO$(gas) (4)
(Boudouard reaction taking mainly place between 500-900 °C[6])

This CO gas can be distributed throughout the batch blanket and will react with the sulfates according to reactions[7,8]:

CO (gas) $+ Na_2SO_4$ (batch) $\rightarrow$ CO_2 (gas) $+ Na_2O$ (batch) $+ SO_2$ (gas) (5)

or at very high partial CO pressures in the batch gas-phase, sulfide may be formed:

$4CO$ (batch) $+ Na_2SO_4$ (batch) $\rightarrow$ $4CO_2$ (gas) $+ Na_2S$ (batch) (6)

ther reactions that may take place are:

$6CO$ (batch) $+ 2Na_2SO_4$ (batch) $\rightarrow$ $6CO_2$ (gas) $+ 2Na_2O$ (batch) $+ S_2$ (gas) (7)

or CO (batch) $+ Na_2SO_4$ (batch) $\rightarrow$ CO_2 (gas) $+ Na_2SO_3$ (8)

The CO gas will also reduce the valence state of polyvalent ions such as ferric iron:

$Fe_2O_3 + CO \rightarrow 2FeO + CO_2$ (9)

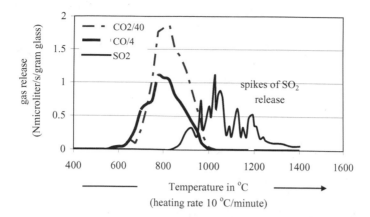

Figure 1   *Evolved gases during the heating of about 100 grams of soda-lime-
silica batch with 1 mass-% sodium sulfate and 0.2 mass-% cokes.
Note that CO and CO_2 curve heights are reduced by a factor 4
respectively factor 40.
There are a few temperature regimes showing SO_2 release peaks.*

Figure 1 shows the evolved gases during heating on laboratory scale a small batch of a soda-lime-silica glass forming raw material with sodium sulfate and cokes in a nitrogen atmosphere.

Figure 1 shows that indeed CO evolution occurs parallel to CO_2 release, but at a lower evolution level in the temperature range between 650 and 1000 °C. SO_2 release starts at about 850-900 °C and continues after the batch is molten. The SO_2-peak at about 900 °C can be related to reaction 1a or 5. The melt with remaining sulfate and melt fragments with sulfide, which have been produced by reaction 6, can mix at lower viscosities and reaction 10 may take place (1050-1250 °C):

$$3Na_2SO_4 \text{ (melt)} + Na_2S \text{ (melt)} \rightarrow 4Na_2O \text{ (melt)} + 4 SO_2 \quad \text{(gas)} \qquad (10)$$

$$\text{or } 3Na_2SO_4 \text{ (melt)} + Na_2S \text{ (melt)} \rightarrow 4Na_2SO_3 \text{ (melt)} \qquad (11)$$

This, results in a SO_2 evolution at about 1050-1250 °C. During the experiment, intense foaming of the melt occurred between 1050-1250 °C due to the evolution of SO_2 gases.

Another reaction that can take place in presence of iron oxides at increasing temperatures in the melt (typically up to 1350 °C):

$$2FeO + Na_2SO_4 \text{ (melt)} \rightarrow Fe_2O_3 \text{ (melt)} + Na_2O \text{ (melt)} + SO_2 \text{ (gas)} \qquad (12)$$

Thus, sulfates oxidize the iron oxides first, reaction 12 seems to occur up to about 1300-1350 °C, this temperature range for reaction 12 has been estimated by the use of thermodynamic calculations for this reaction[11].

The SO_2 release at temperatures in the range of 1000-1300 °C, may be caused by reaction 10 and maybe by reaction 12, and SO_2 bubbles (given by the spikes in figure 1) are released from the melt.

In cases with high carbon contents in the batch, no sulfate may be left after reaction 10 has been completed and this is the case for amber glass melting, most sulfur remains as sulfide.

Figure 2 shows an EGA profile for a float glass batch with sodium sulfate, but without extra cokes addition[6].

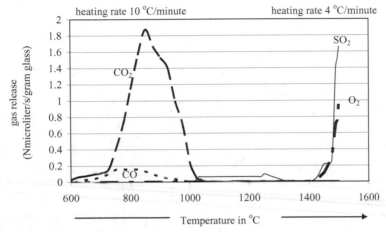

Figure 2   *EGA gas emission profile from soda-lime-silica batch with sulfate addition and no extra cokes (the very moderate CO evolution is caused by small carbon impurity in the sand)*

Figure 2 shows minor CO and SO_2 release at temperatures below 1300 °C. The main SO_2 and oxygen evolution is observed at temperatures > 1450 °C. Most of the sulfate added to the batch did not react with carbon or CO, but decomposes at high temperatures according:

$$SO_3 \text{ (melt)} \rightarrow SO_2 \text{ (gas)} + \tfrac{1}{2} O_2 \text{ (gas)} \tag{13}$$
$$(\text{or } Na_2SO_4 \rightarrow Na_2O \text{ (melt)} + SO_2 \text{ (gas)} + \tfrac{1}{2} O_2 \text{ (gas)})$$

Figure 3 shows another example for a float glass raw material batch with sulfate and only small addition of cokes, it shows 3 SO_2 evolution peaks. The first two peaks (1075 °C and 1225 °C) are associated with reactions, such as 5 and 10 and the last peak at about 1460-1500 °C with reaction 13.
These figures show that gases can be released from the melt at different stages and different temperature intervals, depending on the presence of reducing agents such as cokes in the batch. Probably the grain sizes, homogeneity of the batch and thickness of the batch blankets will influence the gas exchange between batch and atmosphere and the kinetics of the reactions associated with these SO_2 evolution peaks.

For the gas evolution in the range from 1050-1250 °C (mainly SO_2 gas) reaction 10 is very important. The formed gas bubbles may also enhance sand grain dissolution, by their stirring action, as explained by Klouzek et al[8]. The total extend of gas release, depends strongly on the sulfate addition to

the batch and maximum temperature achieved in the melting tank and glass melt oxidation (redox) state. If sufficient sulfate is left in the melt after reaction 10 is completed, the sulfate can decompose at temperatures, typically above 1430 °C, forming SO_2 and O_2 gas bubbles in a low viscous melt. In glass melt flows that do not reach this temperature level, this stage of gas release is missing. Small seeds containing SO_2 and O_2 gas may be re-absorbed during controlled cooling of the melt, however in reduced melts, with mainly SO_2 as fining gas, residual SO_2 bubbles may hardly dissolve upon cooling during the so called secondary fining processes, because of the lack of oxygen to form soluble sulfates from SO_2 and oxygen.

The viscosity of the melt at 1400-1500 °C is in the range of 5 to 10 Pas and at 1100-1200 °C, this viscosity level in the range of about 100-250 Pas. Thus, in a glass melt tank with glass melt trajectories in this tank that do not reach the 1400 °C level, fining can only take place at lower temperatures with high viscosities. Here fining needs much more time (low ascension rate of bubbles) and may be less effective.

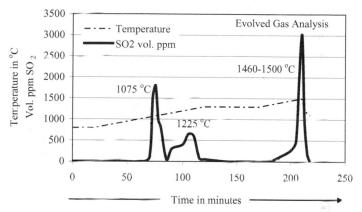

Figure 3 *SO_2 evolution (volume-pm in carrier gas) measured from float glass soda-lime-silica batch as a function of temperature, a small amount of cokes has been added to the batch. Experiment starts in preheated furnace at 800 °C.*

3. BUBBLE ASCENSION IN VISCOUS MELTS
    The velocity of a gas bubble in the molten, viscous glass obeys Stokes'law[1], assuming a density of the bubble to be neglected compared to the glass melt density

$$v_{ascension} = 0.222 \; \rho \cdot g \cdot R^2 / \eta \tag{14}$$

with:

$v_{ascension}$	= ascension velocity of bubble relative to molten glass in $m \cdot s^{-1}$
ρ	= glass melt density in $kg \cdot m^{-3}$
g	= acceleration of gravity force in $m \cdot s^{-2}$
R	= radius of bubble in m
η	= dynamic viscosity in Pa·s

Thus, the same size bubble in a soda-lime-silica type glass melt will ascend more than 1 order of magnitude faster at 1400-1500 °C compared to 1100-1200 °C. For batch compositions with high levels of cokes, most sulfate reacts below 1200 °C with major gas evolution below 1200-1250 °C, here fining needs much more time than at temperatures above 1400 °C and if temperatures are not sufficiently high in the tank fining problems may occur.

Fining at temperatures between 1100-1300 °C by reaction 10, with SO_2 gas release, needs more time than fining by reaction 13 (SO_2 and O_2) at much higher temperatures.

Figure 4 shows the bubble ascension rate in float glass soda-lime-silica melts at different temperatures as a function of bubble diameter.

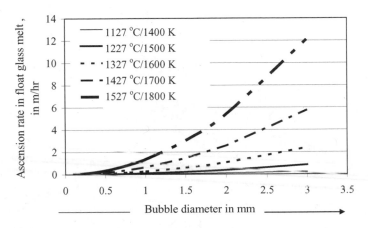

Figure 4   Gas bubble ascension in float glass type melts

4.   SULFUR RETENTION

The sulfur retention after melting and fining of the glass depends on the maximum temperature to which the glass melt has been exposed and the oxidation state of the glass and glass composition (basicity). As will be shown later, it can also depend on the water content in the melt or furnace atmosphere. Increased water levels will decrease SO_3 retention in the glass. The oxidation state can be characterized by the redox number of the batch[9] (Simpson) or the concentration ratio ferrous iron versus total iron in the glass.

For several soda-lime-silica glass compositions the sulfur content after laboratory melting tests has been measured and the results are shown in figure 5. CFe^{2+}/CFe^{total} (in %) is the concentration ratio ferrous iron (CFe^{2+}) versus total iron content (CFe^{total}) in glass. The total iron content $FeO+Fe_2O_3$ is about 0.12 mass-% in most cases. The oxidation state can also be measured by oxygen sensors (based on potentiometry) in the molten glass[10]. The partial pressure of oxygen in equilibrium with the dissolved oxygen in the melt is given on a logarithmic scale on the X-axis (measured at 1400 °C). A high value characterizes an oxidized melt ([10]Log [pO_2/bar] = -1 or -2) and a very low value ([10]Log [pO_2/bar] = -7 or -8) corresponds with a very reduced melt, a typical amber glass.

Figure 5 shows that the sulfur retention decreases with temperature and that the curve: shows a minimum value. At more oxidized conditions, right hand side of the minimum, sulfur preferably dissolves as sulfate and at more reduced conditions as sulfide. At the intermediate oxidation state, sulfur solubility (e.g. a mixture of sulfite and sulfate or dissolved as sulfite: SO_3^{2-}) is very low. A

decrease of the sulfur retention from 0.3 to 0.25 mass-% SO_3 when increasing temperature up to 1500 °C starting from about 1400 °C will lead to 2 liters of SO_2 gas evolution per liter of melt at that temperature level! Part of this evolved gas volume will diffuse in the existing bubbles. Thus, these bubbles will grow (R-value in equation 14 will increase) and will experience an increased ascension velocity in the melt according to equation 14.

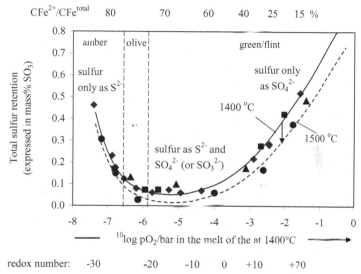

Figure 5   *Sulfur retention (expressed as wt% SO_3) in glass after melting at 1400 or 1500 °C. Dashed line shows interpolation of 1500 °C results, solid line results after melting at 1400°C.*
Base glass composition: 74 mass-% SiO_2, 16 mass-% Na_2O and 10 mass% CaO. Data points from experiments performed at TNO.

5. MECHANISM OF PRIMARY FINING

After <u>almost</u> complete melting of the batch, for soda-lime-silica glass at 1100-1200 °C, the glass is further heated and reactions between sulfide and sulfate rich melt volumes downstream the batch blanket will cause the formation of SO_2 gas, according to reaction 10. In the presence of non-molten particles, such as sand grains, now bubbles containing this SO_2 gas may nucleate and arise, but the viscosity of the melt is still rather high. Intense evolution of SO_2 gases at this stage (reaction 10) may even result in foaming close to the batch blanket tip.

At much higher temperatures, sulfates: SO_3 or SO_4^{2-}, in oxidized melts (in reduced molten glass, sulfur is mainly present as sulfide) can decompose spontaneously in the melt according to reaction 13, and generates oxygen and SO_2-gas in a molar ratio of 1:2.

However, for most soda-lime-silica type glasses in industry, this thermal decomposition reaction requires temperatures above 1400 °C. Thus, a glass melt stream in the melting tank along the relatively cold tank bottom may not achieve this level and this decomposition may not take place in that glass melt trajectory. This may lead to not completely (re)fined glass products with many small

seeds. Especially furnaces with low bottom temperatures and no distinct spring zone (see figure 7) may encounter problems, shown by seeds in the glass product.

The original small bubbles or seeds that are supposed to be removed by an efficient fining process often contain batch gases (e.g. CO_2 from carbonate decomposition) or air species (nitrogen, argon, oxygen). The nitrogen: argon concentration ratio in the bubble in the last case is similar to the nitrogen: argon ratio in air: about 78:0.9.

Due to decomposition of sulfates, the fining gases SO_2 and O_2 (in oxidized melts) or oxygen only as fining gas, in case of antimony fining decomposition of Sb_2O_5 ($Sb_2O_5 \rightarrow Sb_2O_3 + O_2$), these released fining gases first dissolve in the melt, which will become supersaturated with these fining gases:

a. New bubbles may be created in case of the presence of bubble nucleation sites, such as refractory materials or non-molten particles in the glass melt;

b. The dissolved fining gases diffuse to the nearest seeds and bubbles that originally contain CO_2, or air or a combination of both.

c. Only a limited amount of fining gas will escape directly from the glass melt surface into the furnace atmosphere.

The objective of the primary fining process is to generate fining gases for process b. In the case of the presence of non-molten particles in the melt during primary fining, much fining gas is used to create new bubbles (after nucleation) and the efficiency for step b is reduced since less fining gas is available for diffusion into existing seeds. In the worst case, there are still sand particles or other non-molten particles in the melt after the fining reactions are completed. During sand grain dissolution new bubbles may be formed (SO_2 or CO_2 bubbles), since dissolution of sand will decrease the alkalinity of the surrounding melt and reduces the SO_2 and CO_2 solubility. These new bubbles cannot be removed anymore by a primary fining process when all active parts of the fining agent have already reacted (has been decomposed) in the melt.

Conclusion: the sand grain dissolution process should be completed before the melt reaches the sulfate decomposition temperature (in case of sulfate fining) or the antimony fining temperatures.

The diffusion of fining gases into gas inclusions, upon sulfide-sulfate reactions (1100-1250 °C) or sulfate decomposition (> 1400 °C) will lead to the following behavior of these existing seeds and bubbles in the melt:

The internal bubble pressure (P_t) in a bubble in a glass melt tank is given by:

$$P_t = P_o + \rho \cdot g \cdot h + 2\sigma/R \qquad (15)$$

The total amount of gas in the bubble:

$$N_t = 4/3 \ \pi R^3 \cdot P_t /(R_g \cdot T) \qquad (16)$$

with:

P_o = ambient pressure in Pa
ρ = glass melt density in $kg \cdot m^{-3}$
h = glass melt level above bubble in m
σ = surface tension in $N \cdot m^{-1}$
R = actual bubble radius in m
R_g = universal gas constant: 8.31432 $J \cdot mol^{-1} \cdot K^{-1}$

N_t = total moles of gas in a bubble
T = absolute temperature in K

P_t is the summation of all partial pressures (P^{bubble}_i) of the different gas species i within the bubble:

$$P_t = \Sigma P^{bubble}_i \qquad\qquad (17)$$

Diffusion of one or two gas species from the molten glass into a single bubble will lead to and increase of the moles of gas in the bubble and according to equation 16, this will increase the value of R (bubble radius) and will decrease the bubble pressure slightly (equation 15).

Especially when the total internal pressure of the fining gases in the melt produced by decomposition of sulfate or other fining agents, exceeds the P_t level, a continuous diffusion of gases into the bubbles will take place. The glass melt may contain other dissolved gases at relatively high equilibrium pressure, for instance water that may support the bubble growth. If the total equilibrium pressures (P^{eq}_i) of all mobile gas species dissolved in the melt: ΣP^{eq}_i increases above P_t, the fining process starts: the fining-onset. The temperature at which this takes place is called the fining onset temperature.

Thus, the total pressure of the bubble will even decrease by the diffusion of fining gas into the bubble (σ/R decreases in equation 15) and according to equation 17, the partial pressures of the other gases in the bubble will all decrease.

Before fining, the gas species in the bubbles and in the melt are almost in equilibrium, this means that the partial pressure of these gases in the bubble are in equilibrium with the concentration of the same gas species in the melt.

By diffusion of fining gases into the bubble, the partial vapor pressures of the original gases in the bubble will decrease (dilution by fining gas and slightly reduced total bubble pressure) and the equilibrium between gases in bubbles and melt will be disturbed. This will drive diffusion of all gases, that are diluted in the bubble by the fining gases, from the melt into the bubble.

The extra diffusion of fining gases plus other gases from the molten glass into the bubble will lead to extra bubble growth and the bubble ascension in the melt will be enhanced according to equation 14. But, this fining process, as described here will also accommodate the stripping of gases from the melt into the ascending bubbles and therefore reducing the concentrations of these dissolved gases in the melt.

Figure 6 shows the partial pressure of dissolved gases in a float glass melt during heating between 1273 (1000 °C) and 1773 K (1500 °C), calculated from a model described elsewhere[11]. The scale is logarithmic! The figure shows that the fining gas partial pressures (SO_2, O_2) will increase, due to the sulfate decomposition reaction, and that stripping of nitrogen and CO_2 causes the reduction of the partial equilibrium pressures of these gases in the melt.

After efficient primary fining, the concentrations of gases such as nitrogen and CO_2 in the melt and their partial equilibrium pressures have been significantly reduced.

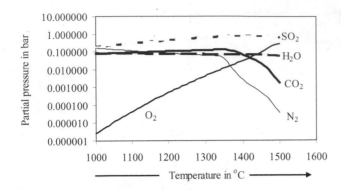

Figure 6   *Change of internal partial pressure of dissolved gases in a float glass melt according to simulation modeling (based on thermodynamics of redox reactions and gas dissolution in these melts) of gas-exchange and gas dissolution in glass melts[11].*

After batch melting, the freshly molten glass is mixed with re-circulating glass melt flowing upstream from the spring zone area of the glass melting tank to the tip of the batch blanket. See figure 7. This re-circulating glass melt flows directly along the glass melt surface from the highest temperature zones of the melt in the tank to the batch and is already well-fined (most bubbles are removed and fining agent has been decomposed). The mixing of freshly molten glass containing un-reacted fining agent with the hot well degassed melt from the hot-spot sections will reduce the concentration of the fining agent (sulfate or sulfate plus sulfide) in the fresh melt. The hot re-circulating melt has been exposed to good fining conditions and there is no excess sulfate anymore in this melt. Therefore an excess of sulfate has been added to the batch to assure sufficient high residual fining agent content in the melt after mixing of the batch melt with re-circulating glass melt. Sulfur concentrations measured for scoop samples taken from the molten glass in the vicinity of the batch blanket in industrial tableware and container glass furnaces show that these concentrations are not very much higher than in the final glass product, this confirms the statement that the sulfate of freshly molten glass is strongly diluted by glass melt already exposed to the highest temperatures.

A very large excess of fining agent is in principal not necessary: Already a decomposition of 0.01 mass-% of SO_3 in the melt will already cause at 1400 °C, a gas evolution of 0.4 m^3 (at this temperature) per m^3 molten glass. Too much sulfate would result in strong foaming.

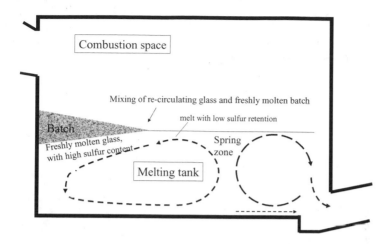

Figure 7   Schematic drawing of main glass melt flows in glass melting tank furnace.

6.   EFFECT OF GAS ATMOSPHERE ON FINING

Exposure of a glass melt to different atmospheres may have an important effect on the fining behavior[12]. Some gases show a high solubility and diffusion coefficient in the silicate melt. Such gases are for instance helium (noble gas molecules with very small diameter) and water (high solubility) and infiltrate easily into the molten glass. Bubbling of the melt[12,13] with these types of gases will even increase the concentrations of these gas species in the melt.

An increasing fraction of the total internal pressure of mobile gases dissolved in the melt can be caused by these gases infiltrating in the melt.

In case of sulfate fining, at equilibrium conditions (e), there will be a relation between sulfate in the melt and the equilibrium partial pressures of SO_2 and O_2 in the bubbles.

$$K_{sulfate} = \frac{aNa_2O \cdot p^e SO_2 \cdot \sqrt{p^e O_2}}{aNa_2SO_4} \tag{18}$$

For constant sodium oxide chemical activity (aNa_2O) and proportionality between sodium sulfate activity (aNa_2SO_4)[14] and SO_3 (sulfate) concentration, we define:

$$K^c_{sulfate}(T) = \frac{p^e SO_2 \cdot \sqrt{p^e O_2}}{[SO_3]} \tag{19}$$

$K^c_{sulfate}(T)$ only depends on glass composition and temperature.

The partial equilibrium pressures of SO_2 and oxygen in the melt determine the maximum concentration of SO_3 (sulfate) in the melt [SO_3] or final sulfate retention at a given temperature T. In the case that the vapor pressures of dissolved SO_2 and O_2 in the melt: $pSO_2+pO_2 > P_t$, the fining agent

will continuously release gas, during primary fining until relation 19 holds. However, when other gases such as water or helium also contribute to gas diffusion to the bubbles, the level of pSO_2+pO_2 may be lower than P_t during fining, but $pSO_2+pO_2+pH_2O + pHe > P_t$. This means that effective fining can already start as soon as $pSO_2+pO_2 > P_t- pH_2O-pHe$, thus at lower temperatures compared to melts without these dissolved gases (He or water vapor).

After fining, the pressure of $pSO_2+pO_2 = P_t- pH_2O-pHe$ and is thus lower than P_t. This means pO_2 and pSO_2 after fining (at same temperature) will be lower in presence of other highly soluble gases than in melts without these gases. According to equation 19, the $[SO_3]$ content at equilibrium with these pO_2 and pSO_2 levels will decrease. This means that more sulfate (assuming the same level of sulfate addition to the batch) has been removed from the melt during fining in melts with helium or more water, and more SO_2 and oxygen gas has been formed in presence of gases such as helium or water vapor. The effect of extra dissolved water in the melt on viscosity is rather low for melt temperatures of about 1400 °C: glass melt exposed to oxygen firing compared to glass molten in air-fired furnace: 2-4 °C lower temperatures for the same viscosity.

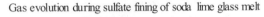

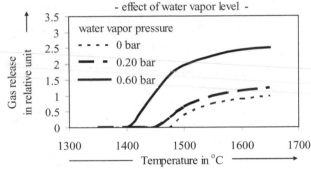

Figure 8   *Gas evolution for glass melt based on the same batch composition, but molten in different atmospheres: dry atmosphere, an atmosphere typically for air-gas fired furnaces (20 vol.-% water vapor) and typical for oxygen-gas firing (60 vol.-% water vapor).*

Conclusions: water vapor and helium gas above the melt or bubbled through the melt will decrease the fining onset temperature, will increase sulfate decomposition and will increase the gas production (extra diffusion of extra gas species and enhanced sulfate decomposition). In practice this has been observed after conversion of air-fired to oxygen-fired glass furnaces, without a decrease of the addition of fining agent, the melt tends to show more foaming, due to extra gas formation during fining. The gas release may be doubled and the fining onset temperature may decrease by more than 20 K, due to the increased water partial pressure above the melt. See figure 8.

Figure 9 shows the measured bubbles sizes during heating and during fining at different temperatures for three float glass melts pretreated in a water containing atmospheres. The average growth rate of 20 bubbles are given per data point. The water content in the melt before the fining experiments has been measured by IR spectroscopy.

The same figure shows curves for the predicted gas release using a gas evolution model described elsewhere[11]. There seems to be a good correlation between the predicted gas evolution (model) and the bubble size average (experiment).

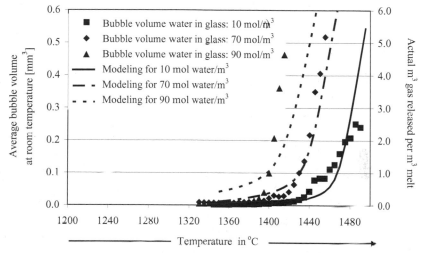

Figure 9   *Observed bubble growth (data points) in soda-lime-silica float glass type melts with sulfate fining and different water contents in the melt. The expected volume of released gases (curves) during fining from a model[11] have also been depicted in the figure.*

Experimental studies[1] with different sulfate contents in a float glass forming batch shows that the fining onset temperature will significantly decrease by increasing the sulfate content. Example: in a typical float glass melt, an increase from 0.36 mass-% SO_3 in the glass before fining to 0.46 mass % will decrease the fining onset temperature with 25-30 °C and increases the gas evolution rate by more than a factor 2 at the same temperature during fining.

7.   SECONDARY FINING

The process of bubble re-absorption during cooling of the melt, often called secondary fining, requires slow cooling. Only gas bubbles, predominantly containing gases with increasing solubility in the melt (SO_2+O_2, CO_2) can be re-absorbed by the melt during slow cooling. Bubbles containing mainly SO_2 may not be dissolved completely, because oxygen gas is required to dissolve SO_2 chemically as sulfate (SO_3/SO_4^{2-}). In that case (bubbles with mainly SO_2, without oxygen), low pressure bubbles/blisters are formed in the final product with deposits caused by dis-proportionation of SO_2 during cooling, forming sulfur and sulfate deposits at the inner bubble wall[15]:

$$3SO_2 \text{ (gas)} + 2Na_2O \text{ (glass)} \rightarrow 2Na_2SO_4(\text{crystal condensates}) + S^0 \text{ (condensate)}   (20)$$

Fast cooling will freeze-in the bubbles, since diffusion of gases from the bubbles into the molten glass will slow down at temperatures below about 1250 °C. Video recording of refining –

secondary fining processes in float glass melts prepared on laboratory scale show that bubbles are re-absorbed by the melt between about 1325 and 1275 °C. At lower temperatures the re-absorption rate becomes very low, due to higher viscosities / decreasing diffusion. Secondary fining will become more effective for melts that have been exposed to an efficient primary fining process delivering a melt with a low retention of dissolved gases (see figure 6).

8. CONCLUSIONS

The primary fining process of glass melts is not limited to the removal of bubbles, but includes also a stripping of dissolved gases from the melt. Efficient stripping of glass melts will reduce the risk of re-boil and enables re-absorption of small bubbles, containing gases with increased solubility at decreasing temperatures.

The fining process, based on sulfate reactions in the melt is rather complex and the process is governed by the total sulfate content in the batch, the redox state or cokes/sulfate ratio, the temperature and the furnace atmosphere. Soda-lime silica batches with sulfates and cokes show sulfur gas (mainly SO_2) release in different temperature ranges. Some SO_2 can already be released from the batch by reactions of sulfate with carbon or CO-gas. The next SO_2 evolution for such batches is observed just after complete batch melting, and this is probably caused by mixing of oxidized (sulfate containing) parts and reduced (sulfide containing fractions) at decreasing viscosity as glass temperature increases. In the case that the addition of cokes is at such level that the content of sulfide formed in the batch is more than 1/3 of the remaining sulfate all sulfate will be reacting at that stage and the melt will hardly retain any sulfate: only sulfide will remain and will lead to an amber color without sulfate decomposition at higher temperatures (> 1400 °C). For cokes lean batches and sulfate fining, the sulfate retained after melting will predominantly decompose at relatively high glass melt temperature forming SO_2 and O_2 for efficient bubble growth and enhanced bubble removal by ascension out of the melt. It is not likely that this fining stage of thermal sulfate decomposition will be met for batches at less oxidized conditions (high amount or cokes or organic materials). Water vapor and helium will reduce the fining onset temperature and will increase the release of gases during fining (in case that fining agent addition to batch has not been changed).

ACKNOWLEDGEMENT:
The author expresses his thanks to Ing. Mathi Rongen from TNO for supporting this study by experimental investigations.

9. LITERATURE REFERENCES

1. Rongen, M.; Beerkens, R.G.C.; Faber, A.; Breeuwer, R.: *Physical Fining Techniques of Glass Melts*. 8th European Society of Glass Science & Technology Conference session B2-Refining and melting II, paper B2-3, 10.-14. September 2006 Sunderland UK
2. Mulfinger, H.O.: *Gasanalytische Verfolgung des Läutervorganges im Tiegel und in der Schmelzwanne*. Glastech. Ber. **49** (1976) no. 10., pp. 232-245
3. Jebsen-Marwedel, H.; Brückner, R.: *Glastechnische Fabrikationsfehler*. 3rd edition (1980) Springer Verlag, Berlin, Heidelberg, New York , Chapter 6 *Schlieren* pp. 309-357.
4. Tammann, G; Oelsen, W.: *Die Reaktionen beim Zusammenschmelzen von Glassätzen*. Z. Anorg. Allg. Chem. **193** (1930) pp. 245-269
5. Flick, C.; Nölle, G.: *Redox conditions during melting of the batch*. Glastech. Ber. Glass Sci. Technol., **68** (1995) no. 3, pp. 81-83
6. Laimböck, P.R.: *Foaming of glass melts*. PhD thesis, Eindhoven University of Technology, June 1998

7. Beerkens, R.G.C.: *Sulphate decomposition and sulphur chemistry in glass melting processes*. Glass Technol. **46** (2005) no. 2, pp. 39-46

8. Simpson, W.; Myers, D.D.: *The redox number concept and its use by the glass technologist*. Glass Technol..**19** (1978) no. 4., pp. 82-85

9. Kloužek, J.; Arkosiová, M; Němec, L.; Cincibusová, P.: *The role of sulphur compounds in glass melting*. Proc. Eighth. Eur. Glass Sci. Technol. Conf. Glass Technol.: Eur. J. Glass Sci. Technol. A, **Vol. 48**, August 2007, no. 4, pp. 176-182

10. Lenhart, A.: Schaeffer, H.A.: Elektrochemische Messung der Sauerstoffaktivität in Glasschmelzen. Glastech. Ber. **58** (1985) nr. 6, pp. 139-147

11. Beerkens, R.G.C.; Van der Schaaf, J.: *Gas Release and Foam Formation During Melting and Fining of Glass*. J. Am. Ceram. Soc. **89** no. 1 (2006), pp. 24-35

12. Kawaguchi, M.; Aiuchi, K.; Kii, Y.; Nishimura, Y.; Aoki, S.: *Helium gas transfer into an E-glass melt with bubbling techniques*. Proceedings XXIst International Congress on Glass, 1.-6. July 2007, Palais des Congrès, Strasbourg, France

13. Beerkens, R.G.C.: *Analysis of advanced and fast fining processes for glass melts*. In Proceedings of the Advances in Fusion and Processing of Glass III July 27-30, 2003 Rochester NY USA, The American Ceramic Society, Westerville. Ceramic Transactions **141** (2004) pp. 3-34

14. Conradt, R.; Scholze, H.: Zur Verdampfung aus Glasschmelzen. Glastech. Ber. **59** (1986) nr. 2, pp. 34-52

15. Golob, H.R.; Swarts, E.L.: *Disproportionation of SO_2 bubbles within Soda-Containing Glasses*. J. Am. Ceram. Soc. **67** (1984) no. 8, pp. 564-567

ALL-ELECTRIC FURNACES FOR HIGH QUALITY DEMANDS IN A WIDE RANGE OF GLASS COMPOSITIONS

Lars Biennek, Harald Jodeit, and Hans-Jürgen Linz
JSJ
Jena-Maua, Germany

ABSTRACT

All-electrically heated furnaces (AEF) are characterized by vertical flow of the melt. The melt itself is continuously and completely covered by a batch layer (cold top). The electric power is released as Joule's heat between the electrodes directly within the glass.

The completely covered melt surface by batch excludes evaporation problems concerning B_2O_3, fluorine, chlorine and so on. Exclusion of evaporation increase the glass quality but also meets environmental regulations to that glass producers get faced more and more of course under consideration of the energy prices.

AEF can be utilised for a very wide range of glass types:

- Soda-lime for tableware but also speciality soda-lime for perfume flacons, STN flat panels and others
- Borosilicate glass type Boro 3.3...4.9, dependend on requirements from side of the final product.
- Opal glass (e.g. tableware and perfume flacons)
- Ophthalmic glass (glass blanks)
- C glass (fibre glass)
- Lead glass
- Non-alkali glass

Each glass type needs a tailor made technology and furnace design. Furnace geometry, type and position of electrodes, choice of the appropriate refractory material and others have a highly important influence on glass quality and furnace campaign. Examples for borosilicate glass (alpha 33 10^{-7} K^{-1}), soda lime and alkali-free glass are announced.

With the development for electric melting of non-conductive alkali-free glass for TFT flat panels and others, JSJ passed an apparently non-crossable border in AEF melting technology. The successful commissioning and operation of the world wide first AEF furnace for non-alkali glass for 2 years is a great step ahead. The development of a special electrically heated system that can be operated independently from the electrical resistivity of the melt surrounding, was the key to success.

AEF is a very good solution for a wide range of glass compositions if focusing to compact construction due to high specific melting capacity, lowest energy consumption, stable and high glass quality, low emissions as well as comfortable operation.

2. PREREQUISITES FOR HIGH QUALITY GLASS

The all electric glass melting process is a little more than just set some electrodes and start glass melting. Especially for producing high quality glass some essential prerequisites have to be considered in designing an all electric furnaces as will be demonstrated by various examples.

2.1 Batch layer

To meet the need of several specific technological demands on each batch and melt there are plenty of varieties realized. Starting from different shapes and depth of the basin, type of electrodes and its arrangements, power distribution phase connections, etc. The first prerequisite for high quality glass is to gain a uniform and particularly stable batch layer. The higher the melting speed of the batch the more challenging the design to keep stable condition and furthermore to guaranty flexibility in melting rate by keeping glass quality.

On the one hand the batch layer has to be stable for thermal insulation but on the other hand the batch composition has to be adapted that those growing bubbles (CO_2, SO_2, NO_x) that flow upward and reach the cold-top batch layer collected from the melt can penetrate.

As a basic disadvantage of AEF over fuel-fired melters the relative small pull range for stable operation occurs, because at lower glass flow rates the batch layer gets thinner and thinner, and the high heat losses do not allow to keep the melting temperatures at a level necessary for sufficient refining. Otherwise, a too high pull generates a very thick batch layer, and raw glass or batch residues are taken with the downward melt convection to worsen the glass quality. Therefore, every AEF owns a designed nominal output and a stable working region of 80…110 %.

In parallel with searching for the working ranges of AEF depending on the specific surface load, also the influence of cullet content on the operation parameters was investigated. A minimum energy consumption was found between 30 and 60 % cullet in the batch-and-cullet mixture. One reason was the decrease of necessary heat supply with growing cullet content, because the share of reaction enthalpy for pure batch became lower. But the opposite tendency, which had to be observed under working conditions in a melter, was the improved heat transport from the melt through the batch layer into the crown. As pure cullet possessed the best radiation condition, the highest heat losses occurred. The sum of heat for loss and for melting presented a more or less severe minimum also depending on cullet size and batch granulometry. It had to be stated that both 100 % cullet and 100 % batch did not allow the stable production of high quality glass under cold-top conditions.

2.2 Temperature and energy fields

The comparable high and homogeneous temperature level in AEF is giving reasons for the possible high specific melting rate. For stable melting condition a very large temperature gradient within the zone of the initial melt below as well as refining zone is necessary. The Figure 1indicates the progression of temperature and density of all process zones. The density of the glass melt starting from the zone of initial melt below the batch blanket until the end of the refining zone is increasing due to chemical transformation from raw material to glass melt. The large density gradient is necessary to stabilize the first processing zones and convection is mainly only possible by temperature and so density gradients within the conditioning zone. The Figure 2: shows the condition of the melt of refining zone taken by a special probe.

Figure 1:

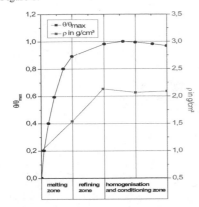

Density and temperature gradient related to the depht of the basin

Figure 2:

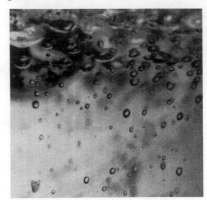

Picture of the refining zone below the batch layer of cold top furnace

3. COMPARISON OF THE THREE MOST SUCCESSFUL SYSTEMS FOR BOROSILICATES 33

Based on model comparison as one example of all-electrically heated furnaces (AEF) to melt 24 t/d of borosilicate glass 3.3 as well based on technological pro and cons due to feedback from the glass factories it shall be demonstrated the essential prerequisites for High Quality Borosilicate Glass.

For this three most successful types of furnaces for borosilicates were chosen. As will be exemplified the major differences are the position of the electrodes, amount of heating circuits, shape and depth of furnace. The vast differences in energy distribution, temperature distribution glass flow and finally glass quality can be demonstrated.

3.1 Characterisation of chosen furnace types

Shelf furnace with Mo plate electrodes is characterized by
- upper basin with about 1.2 m height and a deeper middle channel
- plate electrode rows are arranged near the shelf of upper basin along both length sides
- rectangular melting area with the throat at one smaller front side
- using of one single phase without any division in separate heating zones
- specific melting pull rates of max. 1.3 t/m²d
- specific energy consumptions of 1.15 kWh/kg for the melting part

Figure 3

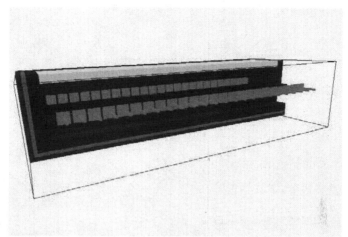

Model of shelf furnace with Mo plate electrodes

Deep square furnace with corner electrodes is characterized by:
- a very deep basin (more than 2 m) with triples of Mo bottom electrodes in each corner situated on increased blocks, protected by Mo ribbon
- using of a 2-phase feeding of the both diagonals with a phase shifting of 90°
- formation of a very thick batch layer in the furnace centre, with the consequence of the big depth for ensuring of quality
- specific melting pull rates of max. 1.6 t/m²d
- specific energy consumptions of 1.20 kWh/kg for the melting part

Figure 4

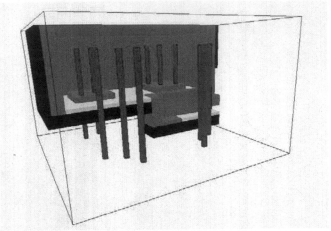

Model of deep square furnace with corner electrodes

Rectangular furnace with two electrode systems and division of them in 3 heating zones is characterized by:

- one bottom electrode system with an electrode arrangement forming 3 squares with 6 diagonals. Each diagonal is a separate heating zone, fed by a 2-phase system
- By means of bottom electrodes 70...80 % of the total power are fed in.
- one horizontal electrode system consisting of 8 horizontal electrode in the same electrical connection as a.m.
- The operation with both systems allows a very high flexibility concerning power distribution in vertical and horizontal direction.
- On strength of the homogeneous power distribution also a homogeneous batch layer is reached.
- specific melting pull rates of max. 1.65 t/m²d
- specific energy consumptions of 1.25 kWh/kg for the melting part

Figure 5

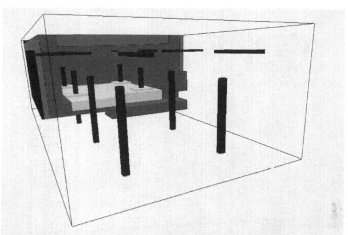

Model of rectangular furnace with two electrode systems and division of them in 3 heating zones

3.2.   RESULTS OF MODELLING

3.2.1.   Shelf furnace with Mo plate electrodes:
(In the cross section one pair of plate electrodes and the deepened channel can be seen)

- The electrical power density shows a rather uniform distribution, however, it is affected by the local variation of electric conductivity (due to temperature differences). (The colours follow the scale of the power density related to its mean value in the basin.)
- The temperature maximum is not close to the electrodes, but in the middle of the basin because sidewalls and electrodes have heat losses. The power concentration is more uniform and not comparable with with rod electrodes. The temperature homogeneity is high, only few differences can be seen.
- The cross convection is very significantly divided in 4 separate convection circuits, but without any fixing by heat sources like electrodes, i.e. a free convection with a high sensitivity against small deviation in heat losses must be accepted
- The *shelf furnace with Mo plate electrodes* presents a very homogenous temperature field but with some downwards deformations of the colder horizon. Especially at the throat side the colder glass is reaching in deeper zones. According to existing experience this behaviour is in agreement with a thicker batch layer at this position with the consequence of too low temperatures on the critical way, expressed in higher losses by glass defects (bubbles) compared with other types, what leads also to a limitation of furnace capacity.

Figure 6

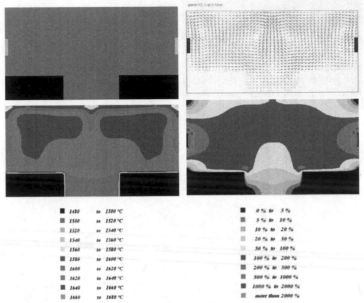

■	*1480*	*to*	*1500 °C*	■	*0 % to 5 %*
■	*1500*	*to*	*1520 °C*	■	*5 % to 10 %*
■	*1520*	*to*	*1540 °C*	■	*10 % to 20 %*
■	*1540*	*to*	*1560 °C*	■	*20 % to 50 %*
■	*1560*	*to*	*1580 °C*	■	*50 % to 100 %*
■	*1580*	*to*	*1600 °C*	■	*100 % to 200 %*
■	*1600*	*to*	*1620 °C*	■	*200 % to 500 %*
■	*1620*	*to*	*1640 °C*	■	*500 % to 1000 %*
■	*1640*	*to*	*1660 °C*	■	*1000 % to 2000 %*
■	*1660*	*to*	*1680 °C*	■	*more than 2000 %*

Results of modelling shelf furnace with Mo plate electrodes

3.2.2 Deep square furnace with corner electrodes
(The cross section is shown through 2 pairs of vertical corner electrodes, i.e. the cross section through the electrodes is situated close to the wall.)

- The power density is concentrated around the molybdenum rods, and between the electrodes, while between electrodes and sidewalls only a minimum of heat is generated.
- Consequently the temperature distribution shows the maximum values at the electrode dips.
- The upward convection is very well stabilised at the electrodes to form two convection circles (upwards in the centre and downwards at the sidewalls). But there is no adjustment of heat generation over the electrode length, and therefore, a secondary convection circle is formed in the lower part between the electrodes.
- The *deep square furnace with corner electrodes* has a rather long distance between the electrodes, and therefore in the centre of the square a critical down flow of cold glass, as well as unmelted batch, may occur. Now one can understand that this furnace needs the deep channel to avoid the short-circuit inflow of batch relicts into the throat. The very low temperatures of 1480°C in a depth of about 0.5 m are in agreement with batch layer measurements, which show a batch layer thickness of 0.5 m.

Figure 7

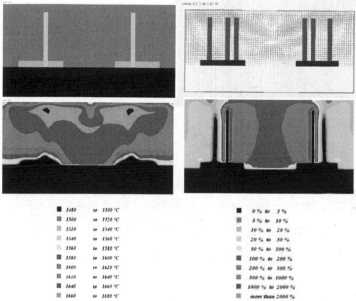

■ 1480	to	1500 °C
■ 1500	to	1520 °C
1520	to	1540 °C
1540	to	1560 °C
1560	to	1580 °C
■ 1580	to	1600 °C
■ 1600	to	1620 °C
1620	to	1640 °C
■ 1640	to	1660 °C
■ 1660	to	1680 °C

■ 0 % to 5 %
5 % to 10 %
10 % to 20 %
20 % to 50 %
50 % to 100 %
■ 100 % to 200 %
200 % to 500 %
500 % to 1000 %
■ 1000 % to 2000 %
■ more than 2000 %

Results of modelling deep square furnace with corner electrodes

3.2.3 Rectangular furnace with two electrode systems:

(Also here a cross section through a couple of vertical electrodes is shown. But this section must be turned by 90° to get the right comparison (two electrode pairs) in comparison with "Deep square furnace". Furthermore it must be noticed that the horizontal electrodes are a little bit outside of the bottom electrode cross section.)

- The main power generation is around the vertical electrodes, but the effect of the two horizontal electrodes can bee seen, too.
- The temperature maximum is distributed more uniform over the furnace.
- Therefore, the convection is stabilised in an ideal way, and the effect of the horizontal electrodes makes an additional forcing the flow.
- The *rectangular furnace with two electrode systems* offers a rather plane batch layer without dramatic cold flow downstream zones as a result of the electrode arrangement with more heating zones.

Figure 8

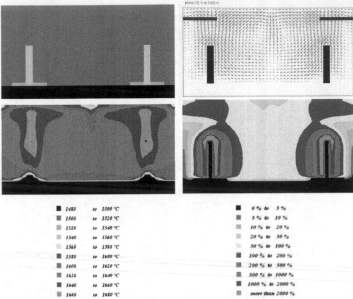

■	1480	to	1500 °C	■	0 % to 5 %
■	1500	to	1520 °C	■	5 % to 10 %
	1520	to	1540 °C		10 % to 20 %
	1540	to	1560 °C		20 % to 50 %
	1560	to	1580 °C		50 % to 100 %
■	1580	to	1600 °C	■	100 % to 200 %
■	1600	to	1620 °C	■	200 % to 500 %
■	1620	to	1640 °C	■	500 % to 1000 %
■	1640	to	1660 °C	■	1000 % to 2000 %
■	1660	to	1680 °C	■	more than 2000 %

Results of modelling rectangular furnace with two electrode systems and division of them in 3 heating zones

3.2.4 Comparison of the critical flow paths

For every point of the melt surface the time to reach the throat entry is calculated for all types, i.e. the individual residence time. The critical flow path characterises the shortest passage of the melting basin. The starting position is very close to the throat wall.

The *shelf furnace with Mo plate electrodes* system is characterized by a free convection with a first cross mixing region at the front side, a flow in channel to the middle of the tank, where again a zone of strong cross mixing must be passed. From there a weak convection flow is going directly to the throat side. The dead time is 13 min 54 sec, in the later comparison to the other types a very short time caused by the a.m. temperature distribution at the throat side.

Additionally to these results the conditions for the area directly in front of throat were investigated regarding the share of glass with shorter residence times as 2 h. In Figure 9 and Figure 10 it can be seen, that very young glass (caused for glass defects) only in a share of 3.8 % < 1 h and 5.1 % smaller 2 h leaves the furnace whereby the heat past indicates that the refining temperature were not reached within this period.

Figure 9

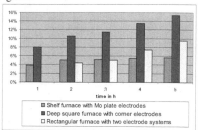

Figure 10

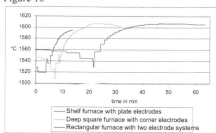

Share of young glass related to residence time   Heat past of first three particles

The analogous conditions for the *deep square furnace with corner electrodes* are shown in the Figure 9 and Figure 10. The intensive mixing of glass particles in the several convection circles is an advantage for high glass homogenisation. The dead time is with 32 min considerably higher (more than two times) due to the long way through the deep channel. But a bigger amount of younger glass is leaving this furnace type.

Again the analogous conditions can be seen for the *rectangular furnace with two electrode systems*. Compared with *deep square furnace with corner electrodes* also here occurs an intensive convection that means a high glass homogeneity. The dead time is with 1 h still longer as the *deep square furnace with corner electrodes* value and very important, young glass with a residence time smaller 1 h is not existing, the young glass share for 2 h is only 4.4 %.

All this together summarized means significant advantages for the *rectangular furnace with two electrode systems* solution, provided that the heat past for the critical way does not lead to a deterioration of the computed conditions.

To clarify such remaining doubts, the temperature history of the three particles on the individual shortest ways through the melting basins is shown in Figure 10. As an explanation for the refining problems with the *shelf furnace with Mo plate electrodes* it is indicated that the temperature maximum is lower and the time is shorter than with the other types. It is a question, if the considerably longer time of *rectangular furnace with two electrode systems* (in comparison with *deep square furnace with corner electrodes*) but with rather the same temperature horizon gives an additional advantage in glass quality.

Generally the shown behaviour gives a good estimation of the refining effect and advantages for adjusting with two electrode systems in two different levels for the furnace.

3.2.5   Conclusions

In the summary of this behaviour a lot of advantages for the *rectangular furnace with two electrode systems* solution according to the forming of batch layer with the consequences for the glass quality influenced by different critical ways are given. The *deep square furnace with corner electrodes* system solves the problem with the big depth of furnace, what is really a necessity as could be shown.

The results of modelling are ratified by real conditions of each furnace type:
 • *shelf furnace with Mo plate electrodes*:
   High sensitivity because of non-fixed convection. The result is fluctuation of glass quality.

- *deep square furnace with corner electrodes:*
  Avoidance of short-circuit of non-refined glass by high depht of the basin and immersed bricks at the wall of throat side

- *Rectangular furnace with two electrode systems*
  Best glass quality and stability but higher expenses for electric heating circuits and higher specific energy consumption

Figure 11

Rectangular furnace with two electrode systems and division of them in 3 heating zones for production of 30 t/d borosilicate glass with expansion 33 10^{-7} K^{-1} (TERMISIL, Poland). Left: Starting of placement of the batch layer after fill melting. Right: Cold-top melting condition with very homogenious batch layer.

4. COMPARISON OF TWO AEF FOR SODA-LIME GLASS
    Based on model comparison as one example of AEF to melt 45 t/d of soda lime as well based on technological pro and cons due to feedback from the glass factories it shall be demonstrated the essential prerequisites for High Quality Soda Lime Glass.
    The melting characteristic of batch material for soda lime glass compared to borosilicates shows a significantly higher melting speed. Therefore local energy overload is connected with destabilisation of the batch layer. Strong convection directly to the batch layer should be avoided to keep homogeneous or rather stable. In the following section the influence of bottom electrodes, top electrodes and combination of both top electrodes and bottom electrodes will be demonstrated

4.1 Hexagonal furnace with bottom electrodes
    Characterized by
- melting surface of 21 m² and the basin is 1.4 m deep
- 12 vertical bottom electrodes. They create a inner circle and an outer circle.
- The electric power supply is realized by 3-phase in star connection, whereby a outer electrode is connected to inner electrode in opposite position.
Typical technological parameters are specific melting rate of 1.9 t/m²d and specific energy consumption of 1.0 kWh/m². The maximum bottom temperature is approx. 1430 °C. The necessary refining temperature for high quality glass of 1480 °C can not be reached due to instabilities of the batch layer.

Figure 12

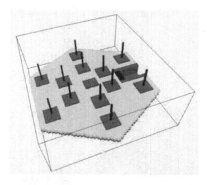

Hexagonal furnace with bottom electrodes

4.2 Dodecagonal furnace with top and vertical bottom electrodes

Characterized by
- 12 top electrodes in the upper part of the basin in 3-phase star connection. Two parallel electrodes are connected to same phase.
- 6 bottom electrodes in the lower part of the basin in three phase star connection.
- Both 3-phase systems (top electrodes and bottom electrodes) are synchronizised
- The melting basin is 1.65 m deep

Typical technological parameters are specific melting rate of 1.9,..2.75 t/m²d and specific energy consumption of 1,0...1 1 kWh/m².

Figure 13

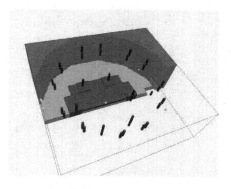

Dodecagonal furnace with top and vertical bottom electrodes

4.3 Results of Modelling

4.3.1 Hexagonal Bottom Electrode furnace

The homogeneity of the energy distribution in cross section of the tank is equal across the total high of the tank, because of the long immersion length of the bottom electrodes. The pictures indicates a maximum of the energy density around the electrodes that cause also a temperature maximum especially at the electrode tips of the inner electrodes, independently of the distance between batch layer and electrode.

This temperature maximum around the electrodes drives the convection from the bottom in direction to the batch layer with high velocity whereby it streams downward between the electrodes as well as the sidewalls.

The convection transports a hot melt directly to the batch layer. Therefore the heat transfer at the locally limited turning round points above each electrode may exceed its limit. The result is destabilisation of the batch layer and increasing heat losses of the open glass batch. In consequence the temperature of the melt may decrease below the refining temperature (assumed once it could be attained).

With increasing distance of electrode tip to batch layer the convection evolves with less limits caused by the longer distance between batch layer and electrode tips as can be seen in the Figure 14 Figure 15 below.

Figure 14

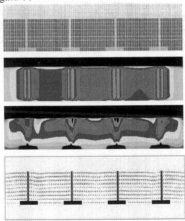

Figure 15

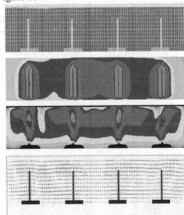

Cross section in direction of throat with low batch layer (1. model, 2. temperature field, 3. energy density field, 4. convection field

Cross section in direction of throat with high batch layer (1. model, 2. temperature field, 3. energy density field, 4. convection field

For quantitative evaluation of the thermal stability it is possible to calculate the excess of energy available directly below the batch layer. This value is characterized by the energy density that is transported to the batch layer by convection and the local need of energy for the melting process and heat losses passing batch the layer.

In Figure 16 it can be seen that the low batch layer horizon cause local energy peaks of approximately 1500 kW/m² compared with a higher level of horizon. The stability of the batch layer is given for the lower level. Under estimation of the same parameters but with a 30 cm higher horizon of batch layer the stability can not be reached because of the energy density peaks exceeds to approximately 2000 kW/m².

Figure 16

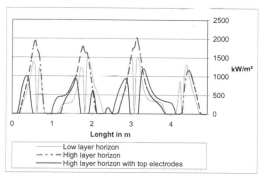

Local excess of energy density of the batch layer

In case of the entire bottom electrodes would have been exchanged by top electrodes and the batch layer is at the higher layer horizon of those that were mentioned above, the positive effect of top electrodes concerning stability of the batch layer is significantly. In comparison of both use of bottom electrodes and use of top electrodes the energy excess of the batch layer can approximately decreased by the half.

4.3.2 Dodecagonal furnace with top and vertical bottom electrodes

In the previous chapter the very positive effect of top electrodes could be demonstrated. This arrangement of top electrodes is hardly realizable, because of the mechanical stability of the electrode holders especially for the inner circle of big furnaces

A combination of top electrodes and bottom electrodes gives a excellent possibility to use stabilizing effect of the top electrodes on the batch layer (decreased upwards convection around the electrodes by approx. 30 % because of Lorenz force and cooling of the batch layer at the areas of ascending melt).

In Figure 17 it can be seen the cross section of the furnace in direction to the throat. The melting rate is 45 t/d (specific melting rate 2.25 t/d). The power supply is exclusively realized by the top electrodes. The energy density at the level of the top electrodes is very homogeneously. The starting strong downward stream in the centre of the furnace is indicated by the temperature field. The bigger the diameter or the higher the pull rate the more intensive the downward stream in the centre will become. This negatively affects the glass quality because the temperature gradient have to be very large for stability of the melting process and sufficient refining. The bottom electrodes can be applied to keep the large temperature gradient also for high pull and big furnaces when using in the rather centric position.

Figure 17

Cross section in direction of throat (1. model, 2. energy density field, 3. temperature field, 4. convection field)

Figure 18

Dodecagonal furnace with top and vertical bottom electrodes for speciality soda lime glass for 53 t/d parfume flacons (HEINZ GLAS, Germany)

Figure 19

Dodecagonal furnace with top and vertical bottom electrodes for fast melting borosilicate alpha 41 10^{-7} K^{-1} (CRISA, MEXICO)

5. AEF FOR ALKALI-FREE GLASS

Glass for the electric industry such as substrates for hard disk and TFT flat panels, for halogen lamps have increasing demand. For this application predominately high quality glass with extraordinary optical properties and low electrical conductivity are demanded. Furthermore in some applications the appearance of alkali reduces the life time of some products like TFT flat panels and halogen lamps. The used glass system is alkali free alumo-earthalkali-borosilicates.

This glass is characterized by high content of boric acid and extreme low alkali content. The melting process of this type of glass is connected with vast technological problems. The production in Pt crucibles is only known for charge production. The melting in oxyfuel heated is connected with numerous problems like evaporation of B_2O_3, necessary high temperatures, chemical aggressiveness against the refractory material, problematic refining and strong crystallization tendency. This leads to high invest costs and high operating costs.

The melting of this glass with AEF offers various technological advantages compared with oxyfuel heated furnaces

- Avoidance of B_2O_3 evaporation
- Realisation of homogeneous temperature field on a high temperature level necessary for providing high glass quality
- Erection of melting plant with melting capacity of 1…12 t/d
- Low invest costs and low operating cost

However for direct electrical heating a minimum level of electrical conductivity have to be achieved.

In the past the AEF-technology for continuous production under cold top could not be realized because direct electric heating with controllable operating voltages is only possible when the

temperature of the melt is higher than 1450 °C. Is there any way to heat up the glass forming melt by a heating system that works independently from the conductivity of the melt and a temperature of 1450 °C can be exceeded then the earth alkalines start to provide a sufficient electrical conducting glass forming melt for direct electrical heating with electrodes.

Figure 20

Model of all electric furnace with combination of top electrodes and Mo heater for alkali free glass

The development of a Mo radiation heater based on an immersion coil for heating water is the key to success. This heating system makes heating of the glass forming melt possible independently from its temperature and as a consequence electrical conductivity. Since the electrical resistivity of the melt is at the same level compared to the high zirconium refractory (ZrO_2 content 96 %) side and bottom electrodes are not recommended. The electric energy would be released within this material in considerable amount connected with local melting off.

The following picture shows an AEF that was erected and commissioned in 2006 for alkali-free glass. In the upper part of the melting basin top electrodes are installed. They give an excellent possibility to install electrodes for direct electric heating without any contact to the refractory material. The major energy supply is given by these electrodes. The minor energy supply is realized by the Mo radiation heater when continuous production is running. When heating up and preparing the melt for direct electric heating the Mo heating system is together with the burner the main power supply.

As mentioned above the expenditures for erection of an AEF melting plant is considerable low compared with an oxyfuel fired furnace. The reason for this is the high specific melting rate in an AEF that is 3.75 times higher and the specific energy consumption is 2.5 times lower under consideration of total melting rate of 10 t/d.

	Melting rate	Specific melting rate	Surface melting basin	Specific energy consumption
Oxyfuel	10 t/d	0.4 to/m²d	25.0 m²	5.0 kWh/kg
AEF		1.5 to/m²d	6.7 m²	2.0 kWh/kg

Beside the higher quality (chemical homogeneity, bubble count, gas contents) that can be produced with an AEF compared with oxyfuel fired furnaces, this technology bears also excellent low energy consumption as well as low invest costs due to low expenditure for the electric heating system and refractory material.

Figure 21

All electric furnace with combination of top electrodes and Mo heater for production of 7 t/d alkali free glass (TELUX, Germany)

6. CONCLUSION

AEF are a excellent solution for a wide range of high quality glass types. Furthermore this melting technology decreases CO_2 emissions, compared to fuel fired furnaces and reduces emissions like boron compounds, fluorine, chlorine and others.

Generally a two electrode system are necessary adjust the temperature about the entire depth of the furnace. Each glass needs a tailor made solution, which can be provided by JSJ based on long time experience. Together with the design we also deliver our well proven key components and the technology as well as know-how.

LABORATORY EXPERIMENTS AND MATHEMATICAL MODELING CAN SOLVE FURNACE OPERATIONAL PROBLEMS

H.P.H. Muijsenberg*, J. Ullrich**, G. Neff***

*Glass Service BV, Maastricht, The Netherlands
**Glass Service Inc., Vsetin, Czech Republic
***Glass Service USA Inc., Stuart, USA
Address for more info: Watermolen 22, 6229 PM; Email: erik.muijsenberg@gsbv.nl

ABSTRACT

Bubbles are frequent defects for glass producers, in most cases it is a small mystery where bubbles are coming from. The main initial source of bubbles is the batch melting. The batch melting reactions release a large amount of CO_2 gas that creates an enormous amount of bubbles. These bubbles need to be refined to produce high quality glass. Glass Service has developed a so called High Temperature Observation (HTO) furnace to observe this initial melting stage and to analyze which raw materials mixture or fining agents deliver the fastest and best seed free glass.

Further Glass Service has developed a special technique and service to quickly analyze composition of gases within the bubble in glass products and based on that may conclude on the source of the defects, as defects do not only come from the batch melting. Furthermore, we can use glass melt modeling and detailed bubble tracing with these models to identify even more closely the possible source of certain defects.

In this article we classify some different bubble sources and bubble formation, as an example. Examples of modeling and bubble gas analyses using a Mass Spectrometer for single bubble categories and bubble source identification are shown. During the presentation of this paper we will show several practical customer examples.

1. INTRODUCTION

Gaseous inclusions in glass are one of the most common defects affecting the glass quality. The position of the bubble in the glass, the bubble size distribution, the gas content, and precipitated inclusions together with the knowledge of interaction mechanisms help the analyst to identify the source of bubble defects. With the help of a dedicated type of a gas mass spectrometer it is possible to analyze the gas phase in gas bubbles in glass. Next to this mathematical modeling helps us to start "bubbles" in the simulation model at certain locations in the tank and trace them to the end of the furnace and verify whether the final bubble composition is the same as with the bubble analysis.

2. HIGH TEMPERATURE OBSERVATION SETUP

The scheme of the high temperature observation method used for batch melting/fining tests is displayed in Figure 1.

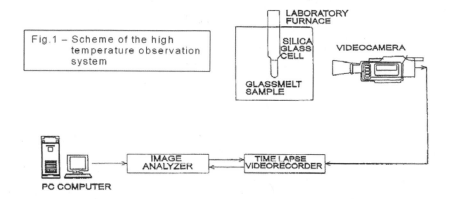

Fig.1 – Scheme of the high temperature observation system

The prepared batch sample contains a given amount of cullet and is placed into the transparent observation crucible made of silica glass. The crucible is then inserted into the observation furnace that is typically preheated at a temperature of 500°C. After heating cnd reaching 1500^{0}C, the isothermal course is kept until complete fining of the glass melt will be achieved.

The course of the test is recorded every 10 seconds within the time sequence, so finally about 1200 images are stored in each test. The video files are created from the saved images showing the entire course of the individual tests as well as a summary file displaying all of the tests simultaneously, for an easy comparison of the batch melting/fining ability.

3. BATCH MELTING TEST

The HTO setup is used to analyze how fast a certain set of raw materials is melting. The transparent silica crucible is filled with raw material and inserted into the HTO furnace. The camera is used to analyze the amount of inhomogeneous area which is a measure of how many seeds are present. In the beginning 100% of the crucible is non transparent, when melting starts the amount of inhomogeneity is decreasing slowly to 0%. When the inhomogeneity reaches 0% it means we have reached perfect seed free glass.

See figure 2 with an example of such an experiment we have done for a container glass composition.

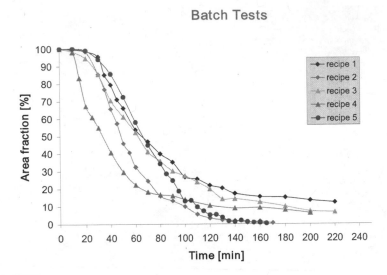

Figure 2. Inhomogeneity melting curve for different container glass batches. 0% represents seed free glass.

In this case recipe 1 was the original batch composition the customer was using. We can see that this recipe shows even after 4 hours melting time still not a seed free glass. It indicates that the fining behavior of this batch was very poor. We than try different recipes or different raw materials to optimize the fining speed. We clearly can see that recipe 2 shows a much faster and better melting curve. For recipe 2 we reach a good glass quality already after just over 2 hours, actually recipe 1 never reached this quality even at 3 hours. Clearly a different recipe really can change the melting curve. Recipe 4 for instance starts melting much more quickly, but still does not reach the best glass quality in the end. Recipe 5 also reaches a very good glass quality after just over 2 hours, however the followed melting curve is slower than recipe 2 which gives less risk to foam formation when temperatures in the melt are still low. So from the above batches recipe 5 would be recommended to the customer.

So the HTO is a very valuable equipment to test different or new raw materials and to find the optimal batch composition.

Figure 3 shows a picture of the melting process after 1 and 2 hours from the experiments, above for recipes 1 till 4.

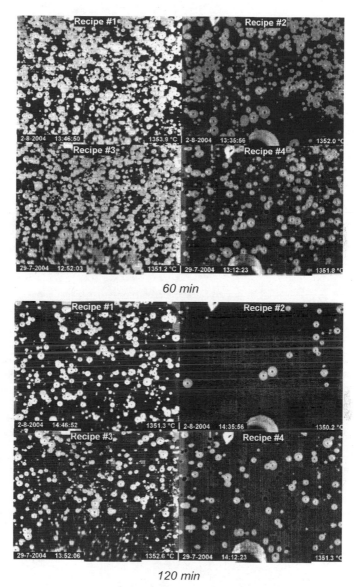

60 min

120 min

Figure 3. snapshot of the melting experiment after 1 and 2 hours melting time. The dark black is the molten glass and the white dots (donuts) are the seeds.

4. DEFECT ANALYSIS AND BUBBLE SOURCE IDENTIFICATION

Knowledge of bubble composition, resulting from mass spectrometric analysis, is the basic step of the systematic approach to the bubble source identification. The basic principle of the identification procedure is studying the available bubble properties as:

- gas composition
- internal pressure
- size distribution, mostly recalculated to equivalent diameter at melting temperature
- presence of condensates
- other accompanying attributes (crystal, stones, cords) in the surrounding glass
- quantity of bubble appearance
- distribution of bubbles in the products

Following the bubble properties, their sorting and evaluation leads to development of extensive knowledge base used in computerized systems for identification of bubble source for industrial tanks. Incorporation of mathematical modeling and bubble tracing procedure should improve predictions not only concerning the glass quality, but also to identify possible bubble sources, its localization in the melting tanks.

We will give some examples in this paper. Note this is just **a selection** there are much more possible sources than given in this paper.

Bubble sources existing in industrial glass furnace. During the melting process, glass melt from batch contains huge amounts of bubbles. Most of them will disappear during the fining course; some new are nucleated later in consequence of various causes and accidents and shortcomings in the process or regular patterns for example corrosion of refractory blocks. Therefore general classification of bubble sources in the glass melting process appears to be quite useful in order to realize the source or places of nucleation. The following categories based on the different mechanisms of bubble generation are proposed:

Bubbles rising from batch decomposition. This is primary source resulting from decomposition of carbonates present in batch. This process generates bubbles containing high CO_2 contents. Besides CO_2 evolution, the (re)fining reactions activated by high temperatures gives oxygen, and SO_2 mostly diffusing into existing bubbles and enhancing their ascending to glass level. Typical bubble composition of melting/refining bubbles is still in the product due to unstable melting process shows the Table 1. Typically small size together with high CO_2, minor content of nitrogen and possible traces of Ar (in range of hundreds ppm) is analyzed in the bubbles.

No.	Dimension [mm]			D.EQ.	Volume	p	Gas composition [%]									Note
	a	b	c	[mm]	[ml]	[kPa]	N2	CO2	O2	Ar	SO2	H2S	COS	CO	H2	
Float	0,20	0,19	0,19	0,19	3,78E-06	17,0	14,4	85,6								deposit
	0,17	0,17	0,17	0,17	2,57E-06	15,0	20,6	79,4								deposit
	0,22	0,20	0,20	0,21	4,61E-06	13,0	19,2	80,8		TR						deposit
	0,19	0,18	0,18	0,18	3,22E-06	14,5	15,7	84,3		TR						deposit
TV	0,18	0,16	0,16	0,17	2,41E-06	26,0	39,4	60,6								
	0,34	0,17	0,17	0,21	5,14E-06	24,0	28,2	71,8								
	0,28	0,15	0,15	0,18	3,30E-06	21,5	42,6	57,4								
	0,40	0,19	0,17	0,23	6,76E-06	28,0	32,5	67,5								

Table 1 - Gas composition of melting/refining bubbles for two industrial glasses - TV panel and Float.

TR - traces; D.EQ - equivalent diameter to spherical bubble

Bubbles rising from nucleation. The reason of bubble generation is supersaturation of the glass melt by gases due to composition or temperature changes and heterogeneous nucleation. Refining gases as SO_2+O_2 in sulfate containing glasses, which are more sensitive to reboiling, and O_2 in case of glasses refined by oxides, generally forms the initial composition of bubbles. Those gases have steep temperature dependence of glass melt solubility, therefore the composition of nucleated bubbles can quickly change upon temperature changes. The new (fresh) bubbles contain high amount of refining gases while the older ones should have significantly less or none of them.

Simultaneous diffusion of CO_2 and N_2 from the melt into these bubbles leads to increasing CO_2 and N_2 contents and give estimation of their age. Low bubble pressure due to SO_2 disproportionate reaction could be often found, too. Important feature is, they frequently contain no argon. Table 2 shows the example of gas composition of bubbles coming from gas reboil. The amber glass is easy predisposed to over saturation.

Sample ID	No.	Dimension [mm]			D.EQ. [mm]	Volume [ml]	p [kPa]	Gas composition [%]							Note
		a	b	c				Ar	CO	CO2	COS	H2S	N2	SO2	
Amber glass	1	0.56	1.24	0.42	0.66	1.53E-04	10.0			9.1	TR.		5.9	85.0	deposit
	2	1.38	1.96	0.38	1.01	5.38E-04	6.5			4.5				95.5	deposit
	3	2.55	0.38	0.56	0.82	2.84E-04	4.0			17.4				82.6	deposit
	4	0.36	1.44	0.70	0.71	1.90E-04	11.0			5.1			0.1	94.8	deposit

Table 2 - Examples of analysis of bubbles rising from temperature reboil in amber glass. Refiner zone. TR – traces; D.EQ - equivalent diameter to spherical bubble

Bubbles rising from electrochemical reaction. The bubble generation can be observed on immersed thermocouples, boosting or earth electrodes, and contact Pt-level gauges, generally on metal parts being in contact with glass melt and forming one of the cell electrode. The second electrode is created usually by refractory. Electrolyte is provided by the glass melt exhibiting conductivity at high temperatures. If both electrodes come into short circuit connection, for example via steel construction of the furnace, a closed galvanic cell is generated and the following reaction can happen on the anode:

$$2O^{2-} - 4e^- \leftrightarrow O_2\,(g,l)$$

Fresh bubbles produced by electrochemical reaction contain mainly oxygen, typically 90 100 vol.% while the older ones should have significantly less or even no oxygen due to its fast dissolution in the cooling glass melt. The source is significant mostly in the low temperature areas of the furnace, feeders; the size of bubbles is relatively big. Table 3 shows typical example of analysis of bubbles coming from electrochemical reaction.

	Dimension [mm]			D.EQ. [mm]	Volume [ml]	p [kPa]	Gas composition [%]										Note
	a	b	c				Ar	CO	CO2	COS	H2	H2O	H2S	N2	O2	SO2	
	3.85	3.55	1.36	2.65	9.73E-03	19.0			0.2					1.7	98.1		
	1.06	1.66	1.28	1.31	1.18E-03	32.5			0.1					0.6	99.3		
	0.20	1.24	0.78	0.58	1.01E-04	24.0	TR.		0.5					1.1	98.4		
	0.60	0.64	0.70	0.64	1.40E-04	25.0	TR.		0.2					1.4	98.4		

Table 3 - Example of analysis of bubbles rising from electrochemical reaction in flint glass generated close to spout.
TR - traces; D.EQ - equivalent diameter to spherical bubble

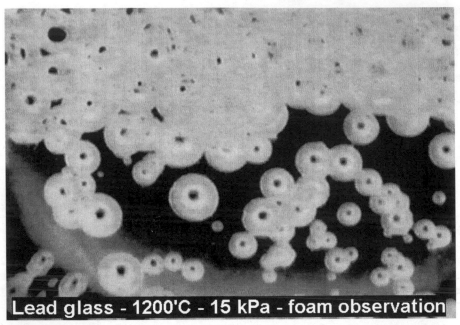

Lead glass - 1200'C - 15 kPa - foam observation

Figure 4 shows how we can monitor bubbles in a foam on the glass surface in the GS High Temperature Observation (HTO) furnace (low pressure – sub atmosphere).

Mechanically formed bubbles in the glass are generated by mechanical interaction between the melt and static or moving parts used in the course of the glass melting process (submerged wall, stirrers, feeding device, or refractory crack etc.). The bubble source or cause is mostly located close to output of the furnace. Bubbles contain air or air residuals, typically argon and N_2/Ar ratio close to that found in ambient air. They do not exhibit significant changes in composition nevertheless sometimes the diffusion process of CO_2 into the bubbles may give an estimation of their age. Table 4 represents the analysis of bubble caused by mechanical entrapment of air or furnace atmosphere in TV panel glass.

No.	Dimension [mm]			D.EQ.	Volume	p	Gas composition [%]										Note
	a	b	c	[mm]	[ml]	[kPa]	N2	CO2	O2	Ar	SO2	H2S	COS	CO	CH4	H2	
1	1,23	0,48	0,23	0,51	7,11E-05	26,5	93,4	5,6		1,0							deposit
2	0,54	0,50	0,51	0,52	7,21E-05	25,0	91,3	7,8		0,9							deposit
3	0,43	0,45	0,42	0,43	4,26E-05	28,0	94,5	4,5		1,0							deposit
4	1,36	0,25	0,12	0,34	2,14E-05	24,5	91,8	7,2		1,0							deposit
5	1,50	0,58	0,28	0,62	1,28E-04	27,0	92,4	6,6		1,0							deposit

Table 4 -   Example of analysis of bubbles originating from feeder atmosphere entrapment in low temperature zone of TV tank - feeder, spout.
TR - traces; D.EQ - equivalent diameter to spherical bubble.

Bubbles released from refractory. They are generated from the open pores in the course of the refractory dissolution. Those bubbles contain generally air residuals in combination with

and N_2/Ar ratio is close or slightly higher to that found in ambient air in dependence on the bubble age. Detectable activity of the source frequently happens when glass is penetrating and directly contacting the refractory layers that are not expected to be in touch with glass. Similar situation could be seen at places where higher temperature or faster glass flow in the melting furnace enhances refractory corrosion. Table 5 displays typical composition of bubbles releasing from refractory in the waist or working end zone of a float tank.

No.	Dimension [mm]			D.EQ.	Volume	p	Gas composition [%]									Note
	a	b	c	[mm]	[ml]	[kPa]	N2	CO2	O2	Ar	SO2	H2S	COS	CO	H2	
1	0.68	0.38	0.36	0.45	4.87E-05	29.0	70.4	28.4		0.6		0.5	0.1			
2	1.00	0.41	0.43	0.56	9.23E-05	27.0	68.2	30.9		0.6		0.2	0.1			
3	0.38	0.25	0.25	0.29	1.24E-05	26.5	68.1	31.0		0.6		0.2	0.1			
4	0.35	0.23	0.23	0.26	9.69E-06	27.5	65.4	33.7		0.5		0.3	0.1			
5	0.23	0.17	0.17	0.19	3.48E-06	27.0	62.8	36.3		0.5		0.2	0.2			
6	0.42	0.25	0.25	0.30	1.36E-05	31.0	59.9	39.4		0.4		0.2	0.1			
7	0.47	0.37	0.53	0.45	4.78E-05	21.5	58.8	40.0		0.6		0.4	0.2			

Table 5 - Example of analysis of bubbles releasing from refractory in a float tank - waist-working end zone.
TR - traces; D.EQ - equivalent diameter to spherical bubble

Next figure 5, shows bubbles that are nucleated from refractory. In fact there are 2 samples (A and B) placed in the High Temperature Observation furnace to study the difference between the amount of bubbles nucleated from the different refractories under identical conditions. The left sample clearly nucleates much less defects per surface and time than the right one. This is an example to show nucleation, but also a method to study quality of refractories in relation to defects.

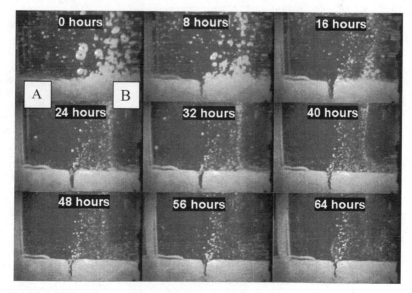

Figure 5. Visual observation of bubbles nucleation from 2 different refractory pieces labeled A and B (left/right)

Mapping furnace defect sources. In case one really wants to know where different bubble "families" in a furnace are coming from it is interesting to analyze about 100 bubbles (or more) from a glass production during a short time period randomly selected. When the results are put into a graph: CO2 concentration versus Argon content for instance or Ar/N_2 ratio, we can distinguish different possible sources for a certain production line. See example of figure 6.

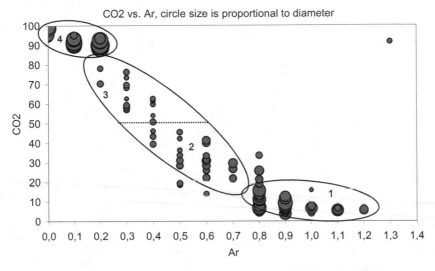

Figure 6. Bubble composition, with different bubble source families plot 1 till 4.

Age of bubbles. Besides the source we also have an ability to indicate from the mars-spec analyses how old the bubbles are. Glass Service has developed special crucible and visual experiments to trace the change of gas composition within the bubble as function of time (with certain temperature course). See figure 7 where we show change of CO_2 and N_2 as function of time (according to measurements and mathematical model at 1300 °C). In figure 6 an original nitrogen bubble, shows how CO_2 diffused in till it a saturation level of above 80%. From this one can conclude (that for this glass composition and temperature at 1300 °C) a bubble with a CO_2 concentration close to 80% is at least 2,5 hours old. In reality it is probably older as temperature in working end and forehearth are lower so the diffusion of CO_2 slows down.

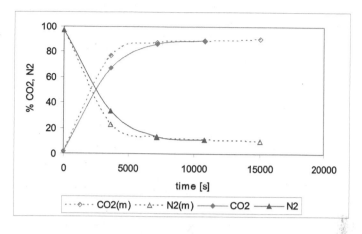

Figure 7. $N_2 \rightarrow N_2 - CO_2$ bubble composition change as function of time at T= 1300 °C
Dashed line with (m) is giving the model results

With this information we can indicate from which area of the glass production the bubble is coming.

5. UTILIZATION OF MATHEMATICAL MODELING IN THE BUBBLE IDENTIFICATION PROCEDURE.

Latest development of mathematical models for furnace simulation allows applying bubble tracing in the melting space using post-models describing the bubble behavior in the molten glass. Despite of their requirements of high accuracy and hardly measurable gas data (solvability, contents in molten glass, diffusion coefficients, reaction equilibrium), they are becoming very attractive because of their high flexibility and limited needs of special operator's knowledge about the "bubble chemistry".

The modeling procedure can be summarized in the following steps:
- Coverage of the expected areas of the source activity by dense network of bubbles of estimated initial composition, sizes or size distribution.
- Tracing of bubbles in the modeled furnace.
- Comparison and/or selection of properties of bubbles reaching the furnace output with appropriate bubble gas analysis.
- Displaying of starting points of bubbles matching the analyzed composition and size gives the estimated area of bubble source location in the real tank.

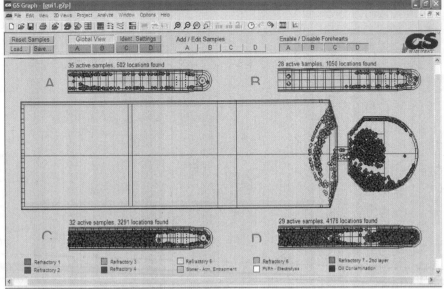

Figure 8 briefly shows the described procedure of bubble source identification using the mathematical simulation model on a TV melting tank example. Each point gives the possible source for an analyzed final gas composition range of a certain bubble type, when compared with the calculated final results.

6. CONCLUSIONS

The combination of visual high temperature observation experiments, mathematical modeling and several thousands of bubble analysis each year give us a special know how in diagnosis of the source of bubbles found in glass products. This quick service is available to every glass producer and supplier to use as diagnosis and receive recommendations how to remove or lower the amount of defects. Next to this Glass Service has developed a software called Expert System I, to help to identify the source of bubbles.

REFERENCES

1. **Kloužek J., Nemec L., Ullrich J.**
Modelling of the Refining Space Working under Reduced Pressure
Glass Science and Technology (Glastechnische Berichte) 73 (11) 129-336 (2000)

APPLICATIONS OF MODEL BASED CONTROL IN FLOAT AND FIBERGLASS

Ron Finch
Advanced Control Solutions Inc.

ABSTRACT

Significant quality and throughput improvement to the glass float and fiber manufacturing process has been demonstrated by a new control method. The new temperature control method utilizes ACSI's predictive-adaptive Model Based Control system. The Model Based Control system continuously adjusts the upstream process to ensure that the flow temperature is at the setpoint regardless of environmental conditions. The system has been field tested in a multitude of installations and has demonstrated the ability to improve temperature stability substantially as compared to conventional PID control. Results of implementing the new control method include improved manufacturing efficiency, product quality, and reduced mass flow temperature recovery time following a temperature setpoint change.

INTRODUCTION

ACSI has successfully combined our glass process knowledge with new innovative technology in our model based control. The combination of ACSI's process expertise and Brainwave® software capability enables our systems to greatly outperform standard PID in all Fiber and Float glass applications. Traditional methods use PID (feed back only) to control variables such as furnace temperature, working end temperature, forehearth temperature or glass level. Our advanced control uses a model to control the process. Once the optimum model is created the controller:

- Predicts control actions required to quickly drive the variable to set point without overshoot

- Adapts to process and production changes for continuous improvement without manual intervention

- Models both feed forward and feedback inputs to provide maximum stability

Model Based Control (*Model Based Control versus PID*)

Model Based Controllers (MBC) are outperforming PID control in glass applications and therefore have become a preferred installation by manufacturers. The MBC quickly responds to process disturbances and reacts swiftly to stabilize temperature variations. A Model Based Controller creates models for each control/process variable and feed forward input. The ideal model then anticipates changes needed to maintain consistent glass temperature. Once the optimum process is modeled, Model Based Control

> • Predicts control actions required to drive the glass temperature to setpoint quickly without overshoot.
>
> •Continuously adapts to process and production changes automatically for better control without loop tuning. (wear and tear of furnace, seasonal changes, etc.)
>
> • Models feed forward inputs and updates control actions to quickly stabilize temperature variation.

Another benefit of Model Based Control is it's consideration of dead time. PID does not understand dead time, which results in oscillation. MBC understands & accounts for it when making adjustments. The

MBC understands what the typical dead time is and waits before continuing with adjustments.

Figure 1: With PID, oscillation occurs before settling at Setpoint

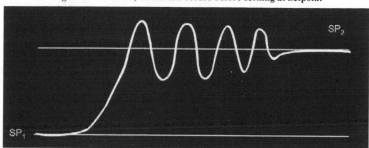

Figure 2: With MBC, the PV reaches SP faster without overshoot.

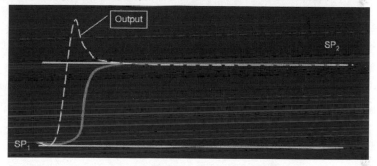

Model Based Control should be applied selectively to avoid high costs and ineffective use.

MBC should be implemented the following areas:

- Loops that are dead time intensive
- Places where it is essential for the control system to understand the relationships between/among loops
- Areas where feed forward signals would increase productivity, decrease process variation
- Anywhere delay time, changing loop dynamics or non-linear responses make for a tough control problem

Overall benefits of Model Based Control:

- Energy Savings
- Greater Overall Stability
- Increased Quality
- Decreased Defects

Float Glass Control

The ACSI MBC is applied to the following loops to determine the effectiveness versus traditional methods:

Melter Temperature control
Canal temperature control
Refiner temperature control
Glass Level control
Block Cooling
Lehr Temperature
Tin Bath Bottom Temperature

The model based control runs in a separate computer that is attached to the control system via Ethernet or Profibus. (Ethernet is preferred) The control system is programmed to allow the model controller to change the outputs to directly control the process. If the model fails the system automatically reverts back to the prior control (normally PID). The only change to the operator interface is an additional mode of control shown (model). The communication interface can be via OPC.

Creating the model requires that the loop is "bumped" through step changes to learn how they response. The Setpoint "bumps" are within limits that do not disturb the running process.

Temperature Control Example

Traditional glass design places a sensor at the exit outlet of each temperature zone to measure glass temperature as it exits the zone. The sensor relays data to the PID controller which adjusts the heat to bring the glass temperature back to set point. As the molten glass travels through each chamber, respective controllers continue to "play catch up".

Also PID controllers do not understand dead time. Dead time is the delay in response to a valve adjustment. Thus PID controllers must be tuned to respond slowly in glass applications.

The ACSI model-based controller is effective in controlling zone temperatures by modeling the existing process. The controller creates models for each control/process variable and feed forward input. These ideal models allow the system to anticipate changes needed to maintain consistent glass temperature.

Figure 3: Temperature and Gas Flow before and after Model Based Control

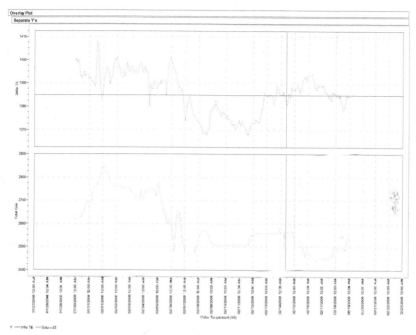

Working End, Canal Temperature and Glass Level Examples

The following trend displays show the affect of model based control when applied to the Working End temperature control.

The first is Working End temperature prior to the application of model based control. The second is using ACSI MBC.

Figure 4: Working End Temperature before MBC

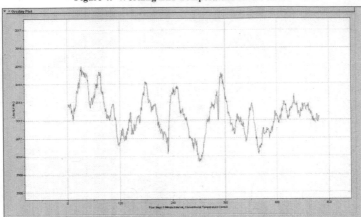

Figure 5: Working End Temperature after MBC

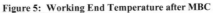

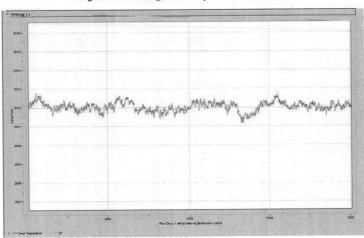

Figure 6: Canal Temperature before MBC

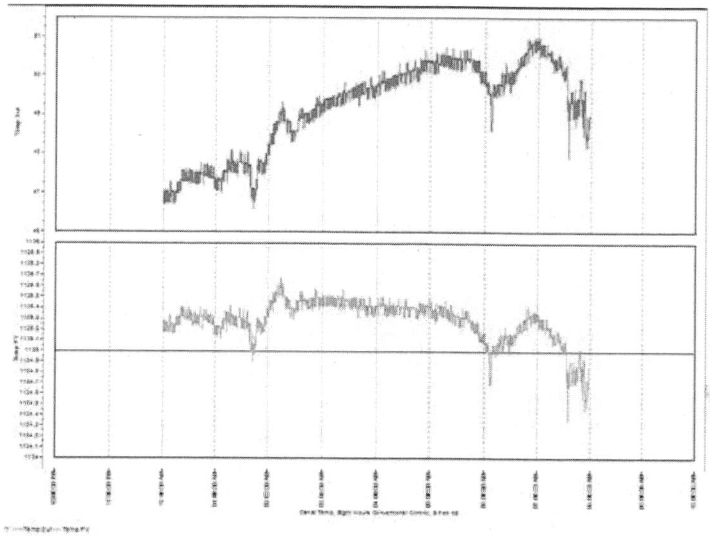

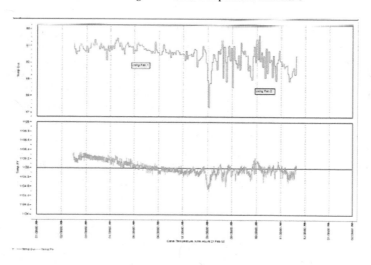

Figure 7: Canal Temperature after MBC

The following images display the affect of model based control when applied to the Glass Level control.

The first is Glass Level prior to the application of model based control. The second is using ACSI MBC.

Figure 8: Glass Level before Optimization

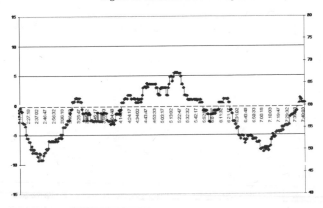

Figure 9: Glass Level after Optimization

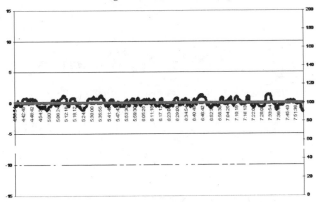

Note: Range shown in "before and after" graphs for both Working End temperature and Glass level are the same parameters

A recent float installation demonstrated our ability to improve the quality of control using a model based controller. The details of the application of this control were as follows:

Leading float glass manufacturers had been using standard PID loops to control glass level and canal

temperature. By incorporating BrainWave® model based control into the existing Melter system, ACSI was able to provide the customer with optimal results that exceeded expectations. BrainWave® technology combines adaptive modeling and predictive control to obtain the best possible results in the least amount of time.

Problems with current solution:
The customer's glass level control, a simple PID loop, was being compromised due to oversized batch logs not melting properly and disturbed convection from the oversized logs cooling the glass beneath the batch blanket. These problems were causing process performance to deteriorate in the forms of poor fuel efficiency and off-quality glass.

The cascaded closed control loop, controlling the Canal temperature, required frequent tuning, produced sluggish response time, and caused an unacceptable variability of the canal temperature.

Results:
A Model Based Control solution provided the customer with measurable advantages both economically and operational.

Benefits
- Increased production
- Energy reduction
- Large reduction in temperature variation
- Noticeable reduction in recovery times
- Improved level control
- Quality and repeatability
- Operator intervention eliminated

Customer Results

- ? 7% reduction of fuel consumption without compromising quality
- 30% reduction of defect density
- Added an additional 10 tons per day without compromising quality
- 50% reduction on error from setpoint in glass level
- Canal temperature variability was reduced by 72%
- Melter Temperature variability was reduced by 67%

Temperature Control for Fiberglass Forehearths

Previous forehearth control systems have relied solely on PID feedback control for glass conditioning to the bushing. ACSI has optimized this control by using advanced model based control to achieve exceptional stability and control right at the point of entry to the bushing. The models use both feedback and feed forward strategies. Customers who have applied the MBC solution have realized significant benefits including improved production efficiency and excellent temperature stability. The solution has been applied to both new and older forehearths using existing control systems with excellent results.

Figure 10: Eight days of forehearth zone control before applying ACSI optimization

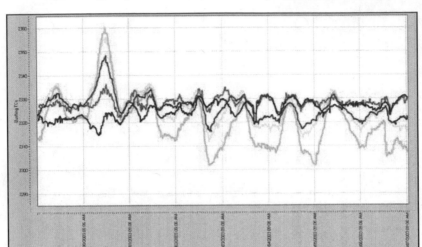

Figure 11: Eight days of forehearth zone control after applying MBC optimization

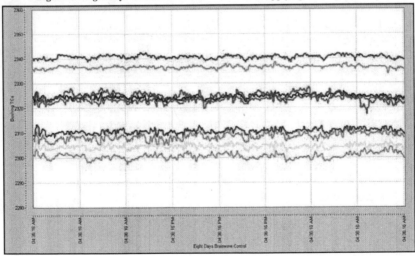

BENEFITS
- Improved Temperature Stability
- Disturbance Rejection
- Reduced Breakage
- Increased Production
- Reduced Energy Usage

CUSTOMER RESULTS

- 65-100% reduction in temperature variation
- 50% less time to achieve stability over PID
- 2% improvement in Pack
- 20% reduction of Bushing breaks per hour
- 2-5% reduction of fuel consumption

LCD GLASS MANUFACTURING FOR A GLOBAL INDUSTRY

Daniel A. Nolet
Corning, Incorporated

ABSTRACT

Corning Display Technologies has seen rapid growth in the past decade along with the global LCD market serving applications in notebook computers, desktop monitors and flat panel televisions. Following some background on Corning, the LCD industry, and market conditions, this paper covers the rapid growth of Corning's business and discusses the special skills and capabilities needed to compete in this market. Unique strategies are needed to understand the global working environment and how to manage a flexible production system with multiple locations to deliver consistent quality to the customer.

CORNING BACKGROUND

Corning Incorporated has a long history (157 years) of manufacturing in the glass industry. Early developments in Pyrex glass for signal lanterns for the railroad industry and in the ribbon machine for high speed production of light bulbs, were followed by many other products in the consumer market (Pyrex cookware, Corelle laminated glass dishware) and in television (CRT panels and funnels for black and white, then color, televisions)[1,2]. Glass manufacturing has been part of the company's genes since its earliest days and today is once again its largest business segment.

In the early 1980's, a small business began to develop in LCD technology. The displays were initially small and poor in quality but worldwide consumer electronics companies saw a future in the technology for larger, portable, higher quality displays. Corning's contribution to the growth of the industry came with a unique process technology, fusion forming, that creates a flat, non-touch surface ideally suited as a substrate in photolithography processing. This process was coupled with unique glass compositions and optical melting technology that provide alkali free, high quality products for use in LCD manufacturing.

Since many display makers have traditionally been located in Asia, initially in Japan, then South Korea and Taiwan, and recently China, Corning has expanded manufacturing globally to develop closer relationships with customers and provide rapid response to customer needs. Figure 1 shows the manufacturing locations for Corning Display Technologies worldwide and it is working in this global environment that will be the main subject of this article.

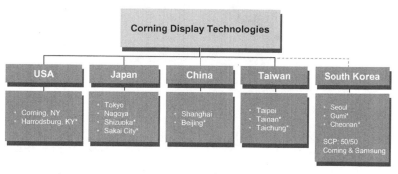

* Manufacturing

Figure 1 – Corning Display Technologies Worldwide Locations

LCD BACKGROUND

Many articles and textbooks now provide excellent references for the structure and design of liquid crystal displays[3,4]. For this application, glass substrates provide the foundation of the display in much the same way that silicon wafers provide the foundation for the computer microchip. Two layers of glass are typically used in a display. The thin film transistor (TFT) side is the most demanding layer with thin film photolithographic processing used to create millions of TFT junctions, four at each pixel location on the display called sub-pixels. The color filter (CF) side holds the pattern of red, green, and blue dots at each sub-pixel location. These two layers are separated by a three to five micron gap spacing which holds the liquid crystal material and clear spacer beads to maintain the gap. Light is transmitted through the display from a backlight module and the liquid crystal, under the influence of current flow controlled by the TFT, acts as a light gate to turn on and off individual sub-pixel locations and create the responsive and colorful moving display images that we see in all our electronics stores today. A simple schematic drawing of the LCD display is shown in Figure 2.

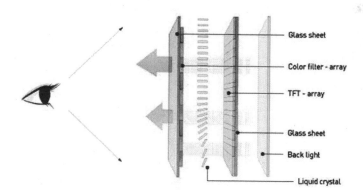

Figure 2 – Typical LCD Structure

For the LCD manufacturer, the key factors driving cost of the final display are both high yield (which results in demanding quality specifications for the substrate as discussed below) and high throughput. Since photolithographic processing represents a very large capital investment, it is critical to size the process to get the highest possible efficiency (throughput) for a given final product size. This has led to a progression of size "generations" in the LCD panel and glass substrate industries. Current manufacturing is heavily focused on Gen 5 and Gen 6 substrates for IT applications and Gen 7 and Gen 8 for television. Gen 10, the world's largest substrate size, is currently under development. This progression of sizes has been facilitated by the inherent scalability of the Corning melting and forming process. Figure 3 shows the progression of sizes over the past few years. Gen 5 was the first size to have one edge longer than one meter. Gen 7 surpassed the two meter edge. And Gen 10 will break the three meter mark.

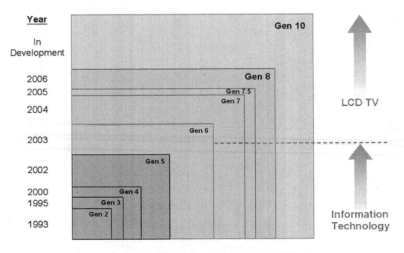

Figure 3 – Substrate Size Generations in the LCD Industry

As a result of this size increase, there has been a roughly exponential growth in the total area and weight of the substrate. This requires an exponential improvement in glass quality to assure consistent yields in customer processes; the same single defect that causes a reject in a Gen 3 piece will also cause a reject in a Gen 8 piece. Figure 4 shows the increase in sheet size matched by improvement in quality over the past 20 years.

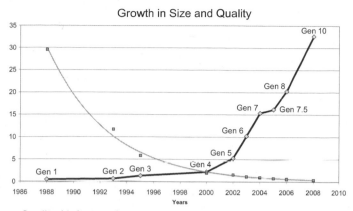

Quality (defect rate) must improve exponentially to maintain yield

Figure 4 – Glass Size by Generation and Quality Improvement

The growth of the LCD market is forecast to continue at 25% CAGR for the next several years. Published data shows the market growth will be heavily influenced by LCD TV expansion as seen in Figure 5 where larger substrate sizes dominate. This also puts stronger pressure on glass quality and glass supply volume.

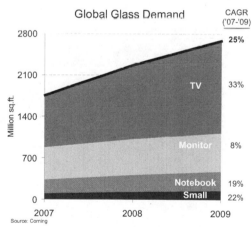

Figure 5 – LCD Market Growth

CORNING CONTRIBUTIONS

The LCD substrate has a very unique set of product requirements to make high yield manufacturing possible. These can be summarized by the triangle of "Clean", "Flat", and "Stable" as seen in Figure 6.

"Clean" in the sense used here means free from defects both internal to the glass sheet (inclusions) and on the surface of the glass (particles). Inclusions larger than 100-300 microns are rejected as they represent possible blockage of sub-pixels as well as possible protrusions on the glass surface. For good Gen 8 sheet production, this means an inclusion level of $4*10^{-4}/in3$ or better is required. For particles, severe requirements on glass chips, stain, scratch, and dirt requires handling processes to be clean room environments as well as completely automated. Corning's glass compositions and optical melting process have proven up to the challenge of delivering to these high performance standards day after day.

"Flat" refers to the surface of the substrate as it is used for photolithography, both in short term ranges (surface height measured over millimeters) and long term range (surface height or sheet thickness over meters of the entire sheet). Short term variation can be measured in nanometers over millimeters which is in scale similar to the height of a 10 centimeter wave over the depth of the Pacific Ocean (>3000 meters). For the longer range, requirements call for <20 microns of thickness deviation over the entire surface of almost 4 square meters. Corning's fusion process has been critical to achieving this degree of flatness without requiring any post-processing, such as polishing, to reduce natural process variation. "Stable" in this context refers to the dimensional stability of the sheet as it goes through the manufacturers thermal processing steps. Most photolithography processes require multiple steps of decreasing temperature to build successive layers on the substrate, with the maximum temperatures reached of approximately 400C. After this process, the TFT side and the CF side (non-thermally processed) must match sub-pixel locations to single digit micron resolution. And this dimensional stability must be predictable and repeatable substrate to substrate, month to month, and across multiple production lines. Here again, Corning has found the fusion process to offer stable thermodynamics of the sheet formation process that allows the customer requirements to be met and the process control to be established to repeat this performance consistently.

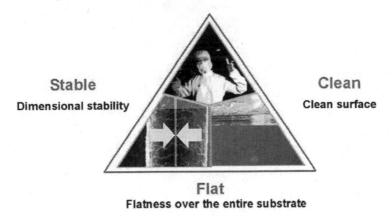

Stable
Dimensional stability

Clean
Clean surface

Flat
Flatness over the entire substrate

Figure 6 – LCD Product Specification Space

These difficult product requirements are quite unique in the glass industry. Corning's glasses are required to be alkali free to prevent contamination of the liquid crystal material through diffusion. The latest generation of glass is also heavy metal free to promote environmental responsibility. The melting process has evolved from Corning history in optical glasses, where high quality and flexibility for production have been important criteria over time. Compared to other glasses in consumer or industrial applications, the quality requirements of LCD substrates (for inclusions as an example) are more than four orders of magnitude more difficult. The final contribution Corning brings is the unique fusion forming process, an invention from the 1960's that has found a new life in the Display Technologies business. Like the invention of the ribbon machine enabling high speed manufacturing of light bulbs, the fusion process has enabled LCD manufacturing to achieve semiconductor industry requirements, allowing millions of transistors to be manufactured at the submicron level on a glass sheet more than two meters (and soon more than three meters) in each dimension.

THE GLOBAL WORKING ENVIRONMENT

Let us now return to the unique challenges presented by working in an industry, as many industries are today, spread across multiple locations, time zones, languages, and cultures. We (Corning Display Technologies) deal first and foremost in an industry that is tightly linked to consumer electronics (notebook computers, personal computers, televisions, and a variety of handheld devices). This brings a pressure of fast cycle time for innovation. We have found the life cycle for a glass composition to be less than 10 years; in the glass industry, where we have made soda lime compositions for hundreds of years, this is a change of thinking at a fundamental level. It forces a continuous balance of innovation (working on the next generation or generations of products) while at the same time working to control and optimize the manufacturing of the present generation. We have approached this dilemma by using our multiple locations to advantage by focusing on manufacturing excellence in most plants, while developing new products in a few locations. This allows us to turn Corning's many systems worldwide into a strength as opposed to a liability.

Global business, as all who participate as a vendor or a customer know, requires personal contact. This means travel and lots of it. It also means learning to listen in different cultures as colleagues try to share their experiences or customers share their concerns and problems. With a global business, we can usually work with a customer in their native language, helping to make quality or supply discussions go much more smoothly. Our internal discussions then must bear the burden of translating across those cultural barriers. Corning has several programs for cross cultural training and awareness, as well as a "Cultural Navigator" to assist in this area.

With a global network of locations, we often find ourselves working long hours just to communicate. Evening or morning conference calls and emails are the norm, whether you are located in the US or in Asia. However, it is also possible to turn this problem into a strength, when you can define a problem for your overseas colleague, go to bed, and then wake up to a solution they have generated during their working hours. By diligent communications (keep the phone calls and email coming), it is possible to work in a more synergistic way to solve problems and meet new challenges.

Finally, the global diversity puts a lot of strain on our people. We struggle to develop leaders and technical experts in new cultures and locations, and give them opportunities to work in other locations to experience the broader Corning family and the international issues facing us. Corning has created a development approach for Global Emerging Leaders to encourage and groom the next generation of business leadership and a career ladder for Subject Matter Experts to encourage technically oriented problem solvers to grow in this skill to become the next generation of technology leaders.

Corning Display Technologies has grown enormously from that small initial LCD business in the 1980's. The challenges of working in a global industry are unique but can be turned into opportunities for faster response to customers, future business and career growth, and participation in a world changing industry.

REFERENCES
1. The Generations of Corning; Life and Times of a Global Corporation by Davis Dyer and Daniel Gross, Oxford University Press, 2001
2. Corning and the Craft of Innovation by Margaret Graham and Alec Shuldiner, Oxford University Press, 2001
3. Active Matrix Liquid Crystal Displays; Fundamentals and Applications by Willem den Boer, Newnes, 2005
4. Liquid Gold; The Story of Liquid Crystal Displays and the Creation of an Industry by Joseph Castellano, World Scientific Publishing Co., 2005

Refractories

BUYING REFRACTORIES IN CHINA

Charles E. Semler
Refractories Consultant
Phoenix, Arizona USA

INTRODUCTION

The "World of Refractories" has dramatically changed over the last decade, with the impact of numerous factors such as corporate consolidations, the internet, and China. Consolidations have resulted in many corporate changes including the elimination of some well-known brands and products, downsizing of staff by refractory users and producers, resulting in increased work per capita, and changes in purchasing practices by the users, e.g. to reduce costs and benefit from the economy of scale. The internet has greatly increased the number of candidate refractory suppliers and brokers available to the users, and thus has enabled changes in buying practices with globalization of the refractory marketplace, and the opportunity for instant communications and online purchasing events which can prompt low bids from unknown, unproven companies anywhere in the world. If price is the main or only criterion for an order, and especially if a refractory technologist is not involved, there is increased potential for problems related to the quality of refractories, which directly affect their service life.

And possibly the biggest change in the refractories world has been caused by China, which is now by far the worlds largest user, manufacturer, and exporter of refractories. Figure 1 shows the increasing production of steel and refractories in China since 2000. The production of steel is shown because the steel industry is by far the largest user of refractories in China (and the world), utilizing about 75% of the annual refractory production in China, and thus is the major market force that drives refractory demand. Since 2000, the tonnage of steel and refractories produced by China has increased 284% and 434%, respectively. In 2007 the tonnage of refractories produced in China was 33.9 million metric tons, which is 14X more than the United States and 31X that of Japan. And the refractory exports available from China have captured the attention of refractory users worldwide, because of their significantly lower price. Figure 2 illustrates the average price per ton in China, compared with the United States and Japan; the overall annual average price per ton (market value/tons produced) for Chinese refractories is 2.3X and 3.3X less than the United States and Japan, respectively, which indicates the general level of cost savings that are possible. But the realization of the cost savings in buying Chinese refractories must involve planning for, and dealing with a variety of issues, including the language barrier, culture, specifications, test methods, production schedule, variable quality, packing/palletizing, truck and train transport in China, container availability, and shipping, to name a few. This paper briefly addresses some of these issues for glass refractories.

BACKGROUND ABOUT CHINA

My involvement with the refractories industry in China began with a three week visit in 1983. Since then I have returned to China 21 times, to visit plants, labs, institutes, mines, etc., in more than 30 cities. So this talk/paper draws heavily upon my personal observations and experiences in China, including the inspection and testing of imported Chinese refractories.

"A changing China is changing the world" (1). To quote a McDonalds Vice President about their business in China, "The opportunities in China are endless, but not effortless" (2). As stated by a Chinese business executive, "You need connections in China. This country works on connections. If you are a foreigner, then you need people who will teach you the strange customs and mindset of the Chinese people" (3), and help you navigate through the system. Considering refractories, and other well publicized products (toothpaste, toys, chocolate, etc.), "Sometimes it is a shock to discover how poor the quality processes are (in China). It is very common that the goods you receive are not exactly what you ordered, either because the factory can't deliver or because the definition of the product is not clear enough" (4). So to take full advantage of the benefits offered by China, it is important to involve knowledgeable people to do the necessary due diligence, and deal with the different cultural issues and business practices. "A major factor in sourcing refractories from China is finding the right partner(s), and because China is a massive country, with extreme variation in price, quality, and production ability, this can be a daunting task. For example, there are no Yellow Pages or industry guides, ISO certification doesn't have the same significance as in the West, suppliers may not be able to, or want to, conform to foreign specifications, and independent lab results may be biased" (5). There are cost benefits to be gained by buying refractories from China, but it is necessary to understand that the ground rules are different, and that business as usual will probably not yield the best results. In most cases it is preferable to involve a native Chinese partner(s), i.e., broker, trading company, professor, etc., although some companies have proceeded to develop the experience to deal with suppliers directly. When buying in China it is wise to bear in mind the old adage, "You get what you pay for", and take whatever steps are necessary to reduce or minimize the possible problems.

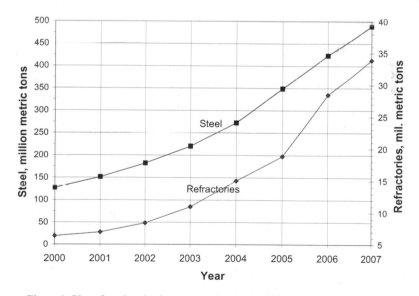

Figure 1. Plot of steel and refractory production in China, 2000 to 2007

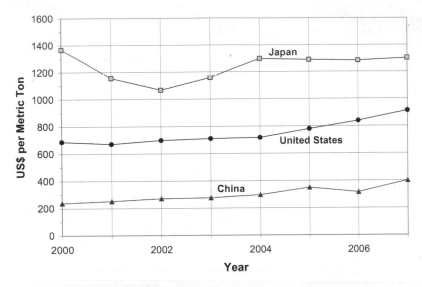

Figure 2. Plot showing that the overall average price per ton of refractories in China
is 2.3X and 3.3X less than the U.S. and Japan, respectively.

CHINESE REFRACTORIES INDUSTRY - GENERAL

It is estimated that there are up to 2000 refractory companies in China, including many small companies. The Chinese government has initiated efforts to reduce the number of companies that are borderline or worse, i.e., those that do not consistently meet quality standards. Ongoing privatization, with elimination of government subsidy, is contributing to the elimination of the weak companies. The Association of Chinese Refractories Companies (ACRC) lists about 190 member companies, which is probably a reasonable estimate of the most reputable refractory companies in China.

The following general observations apply to many refractory companies in China, and they illustrate a few examples of issues that may need to be considered when buying refractories in China:

a. Many companies are not yet computerized, so scheduling, raw material monitoring, production records, quality control, inventory, test reports, etc., are done manually on paper.
b. The chemistry of incoming raw materials may not be consistently monitored.
c. Crushing, grinding, and mixing may involve steel media, which can contaminate materials and batches.
d. Batching may involve beam scales that permit more variation and human error than digital scales.

e. Tunnel kiln cars are tracked using individual hand-written cards.

f. Except for porosity/density, crushing strength, and some chemical analyses, plant labs tend to be little used.

g. The production capacity of most refractory product types is being increased.

h. The number of employees is being reduced, and companies in rural locations are having trouble attracting and retaining employees because of the opportunities and higher salaries in urban centers.

i. Electric power outages can disrupt the control and continuity of production.

j. They are accustomed to doing business with Chinese customers, who seem to be more tolerant of deviations from specifications and quality standards. They will try to follow their usual practices for foreign customers, if allowed.

k. Chinese plants will try to delay product inspection and testing as long as possible, so that there is not enough time to re-manufacture off-quality products by the specified due date.

l. Train shipments by the government receive priority in scheduling.

m. It appears that ISO certification can be bought in China.

REFRACTORIES FOR GLASS – GENERAL

The consumption rate for glass refractories is about 5 kg/ton of glass produced (6). That figure has progressively decreased over the decades as the refractory campaign life has increased, e.g., 6 yrs. in 1980 to 11 yrs. (container glass) and 13 yrs. (flat glass) now (7). The glass industry consumes about 0.5 million metric tons (mmt) of refractories per year; roughly 45% of that amount is fusion-cast products, followed in order by basic, aluminosilicate, silica, and others (zircon, chromia, tin oxide, etc.) (6). Glass contact refractories exhibit the most wear (and premature failure), with a rank-order of (highest to lowest): sidewalls, bottom, throat, regenerator, and superstructure (7).

Refractories are normally the largest expense for a furnace build, and for the repairs (6), which is an important reason why the lower price of Chinese glass refractories has gained the attention of refractory users worldwide. But given the ongoing trend of using glass refractories that are more durable, and capable of providing ever longer service life, it must be recognized that the cheaper Chinese refractories may not allow the stringent performance/cost requirements to be met, and thus shouldn't be considered. But also there are cases where detailed economic analysis of cost effectiveness (refractory cost per ton of glass produced), coupled with careful specification, inspection, and testing of production in China, to insure the refractories are consistent good quality, might justify the usage. And as noted below, there are several glass refractory producers in China, who are using foreign technology, and rank as top-tier, world-class suppliers, whereby the issues related to cost effectiveness and quality may be less of an issue, and if their pricing is good, it is possible to get refractories that are equivalent to foreign-made products at a lower price.

So glass producers now have a much wider choice of refractory sources, but it is now more important than ever to oversee production, inspect products, and do testing of refractories to insure that they will give the performance required for cost-effective, long-term glass production (8).

REFRACTORIES FOR GLASS – CHINA

1. Product Types and Comparison

The main types of glass refractories manufactured in China include fusion-cast, silica, aluminosilicate, basic, and unshaped. The fusion-cast types include AZS (33% ZrO_2), AZS (36% ZrO_2), AZS (41% ZrO_2), AZSC (with added chromia), and high zirconia (>90% ZrO_2). The current capacity for production of fusion-cast refractories in China is 80,000 metric tons. Other Chinese fusion-cast products include super-low exudation AZS (17% ZrO_2), AZS cruciform checkers, alpha-beta alumina, some beta alumina.

There are two joint venture companies in China that are using foreign technology to make fusion-cast products, who are considered top-tier producers; also there are plants that have foreign investment but no sharing of technology, who are considered 2nd tier producers. And there are many Chinese plants that are lower ranked. So when buying refractories from China, it is advisable to know the producer(s), and their relative ranking, to help understand and evaluate the pricing, and guide the level of effort that will be needed to insure that the refractories will be consistent, acceptable quality. A comparison of several Chinese (9) and foreign AZS products is shown in Table 1.

Table 1

General Comparison of Chinese and Foreign Fusion-Cast AZS Products

	Chemical Composition			Bulk Density	Porosity
	ZrO_2	SiO_2	Na_2O	gm/cc	%
China Std. AZS 33 (9)	32-36	16	1.5	≥3.7	≤2
China A AZS 33	33-34	---	---	3.48-3.6	---
China B AZS 33 (10)	32.5	16.2	1.28	3.73	---
Foreign 1-33	33	13.5	1.3	3.72	---
Foreign 2-33	34	15	<2	3.7	<1
China Std. AZS 36	35-40	14	1.6	≥3.75	≤1.5
China A AZS 36	36-37	13	1.3	3.56-3.67	---
Foreign 1 – 36	35	12	1.9	3.8	---
Foreign 2 – 36	36.4	13.3	<2	3.78	<1
China Std. AZS 41	40-44	13	1.3	≥3.9	≤1.3
China A AZS 41	39-41	---	---	3.81-3.86	---
China B AZS	41.25	12.2	1.3	4.0	0.6
Foreign 1 – 41	41	12	1.0	3.97	---
Foreign 2 – 41	39.5	13	1.1	3.93	<1

The data in Table 1 indicate that Chinese AZS refractories can have lower density and higher porosity than foreign-made AZS refractories. And, it is noteworthy that for China manufacturer A, the reported density of their AZS refractories, in all cases is less than the Chinese standard. In a general way, this simple observation documents the reason why a buyer should have input from a refractory specialist, when buying from an unproven company in China (or anywhere), rather than making a purchase decision based only on price and availability.

2. Improvement of Chinese Refractories

Figure 3 (10) provides a visual indication of the progressive improvement of glass tank refractories, design, and operations in China, which have resulted in an ongoing increase in campaign life. These data provide documentation that the improvement of glass furnace refractories proceeded on a continuous basis up to 1998, and my experience has shown that the topical R&D, and increased knowledge of and application of science and technology, have continued to yield improved refractories of all types to this day.

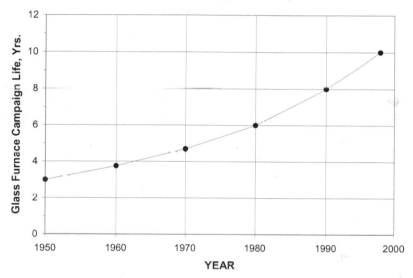

Figure 3. Plot showing the ongoing improvement in glass furnace campaign life in China from 1950 to 1998 (10).

3. Inspection and Testing of Chinese AZS Blocks

The opportunity to observe the production, and inspect, AZS glass furnace blocks in China this year (2008), provided valuable insight into the purchase of refractories in China. Knowing that there could be concerns due to quality, dimensions, squareness, surface condition, internal defects, properties, chemistry, etc., a plan was developed, including sonic testing, to confirm that the refractories were acceptable, before shipment from China. Some people indicated that sonic testing was not useful for fusion-cast AZS blocks, but knowing that it was very important to make a relative comparison of blocks, and have the opportunity to check the internal structure non-destructively, sonic testing was included anyway. Selected highlights of the experience are summarized below, to indicate some of the factors that directly affect the quality of fusion-cast AZS blocks.

3.1 Batching of Raw Materials
    Preparation of most of the AZS raw material batches for the order, which were melted in an electric arc furnace, was monitored. It was learned that the raw materials (alumina, zircon, baddeleyite, and soda) should be better controlled as to chemistry, particle size, and moisture. Instead of using their manual beam balance, a digital platform balance provided much better consistency in weighing each raw material for each batch.

3.2 Addition of Recycled AZS to the Batches
    The standard sized pieces of recycled AZS which were added to each molten batch were roughly 8"-10". These pieces are too big for efficient melting, and would contribute to regions of non-homogeneous microstructure. Upon request, the manufacturer reduced the size of the recycled AZS pieces to <4".

3.3 Oxidation of the AZS Melt
    Oxidation of the AZS melt was done manually, without any sensors or gauges, hence this aspect of the process is subject to variation, and thus a possible issue for the consistency of product quality.

3.4 Cooling of the Poured AZS Ingots
    The molten material was poured into molds, which were then moved to an open bay area for cooling. The area was not heated/cooled so the conditions/rate of cooling were governed by the ambient conditions. Workers indicated that the faster cooling during cold/cool weather resulted in more cracking of the AZS blocks. Clearly controlled cooling of the molds could help reduce cracking of the blocks, and contribute to improved quality, and process yield.

3.5 Stripping of Molds
    Stripping of the molds to release the AZS blocks was done manually. The workers did the job aggressively with a sledge hammer, which frequently resulted in damage or breakage of AZS blocks, reducing the yield.

3.6 Inspection of AZS Blocks in Layup
    Inspection of the AZS tank and throat blocks, arranged according to their furnace position, showed that the blocks fit together well. Dimensions and squareness were within the specifications. Numerous blocks showed spider-web cracking, which was within the specifications. Several open cracks of limited extent were observed, but they were not in glass contact areas, and thus considered tolerable.

3.7 Non-Destructive Sonic Testing of AZS Blocks
    Sonic measurements were made through the edge and middle thickness (300 mm /11.8") of 25 tank blocks; the readings through both the edge and middle were relatively consistent. The average results were 21,533 and 19,799 ft/sec for the edge and middle of the blocks, respectively. None of the readings on any block indicated a significant differential in the sonic velocity, but the differential (21,533 vs. 19,799 ft/sec) indicated that the middle region of the blocks was lower density and maybe higher porosity than the edge. This structure difference is likely a relic of the casting/cooling process.

Five blocks were sonic tested to determine the effect of a visible casting scar in the center of the top face. Sonic readings through the 120 mm (4.7") thickness of the test blocks are shown below:

| | Sonic Velocity, ft/sec | | Edge/Center |
	Edge	Center	Ratio
No Scar	20,721	20,399	1.02
No Scar	20,505	20,190	1.02
Top Scar	21,339	19,685	1.08
Top Scar	20,942	18748	1.12
Top Scar	20,997	19,299	1.09

One electrode block was rejected because of a visible open crack which did not meet specifications. Two other electrode blocks showed possible cracks which had been filled with a mortar filler; these two blocks were sonic tested to compare the readings through cracked (the sonic pulses were transmitted perpendicular to the crack plane) and uncracked material, with the following results.

| | Sonic Velocity, ft/sec | | |
| | Good | Cracked | Good/Cracked |
	Material	Material	Ratio
Visible Crack	21,584	19,764	1.09
Visible Crack	22,472	20,252	1.11

Sonic testing was found to be an effective method of checking the consistency of AZS blocks, comparing their internal structure, and detecting internal cracks/voids. In general, it was seen that as sonic readings increase from 20,000 ft/sec, they indicate material of increasing integrity, and vice-versa, as the sonic readings decrease from 20,000 ft/sec, the integrity of the material decreases. But more experience is needed to refine the technique and improve the capability of interpreting the results.

CONCLUDING COMMENTS

These are definitely exciting, unique times in the refractories industry in China, and the world. The effect of China is global, impacting all segments of the refractories field. It is true that significant cost savings can be realized by buying Chinese refractories. But to actually realize the savings, careful planning and oversight are essential. Without an awareness of the risks, and a plan to avoid those risks, a good initial price can become a very bad (and inflated) price, for a variety of reasons; several examples are: (a) off-quality products being received and installed, which result in a furnace failure, excessive maintenance, or other problems, (b) delays in production of acceptable quality refractories, or receipt of pieces that don't meet dimensional tolerances, which create the need to obtain selected pieces on the open market at a high price, and (c) receipt of refractories that have a high NORM (naturally occurring radioactive materials) count.

This paper reports a comparison of properties of Chinese and foreign-made AZS refractories, and the results provide a general indication of why precautions should be taken to protect your investment in refractories purchased from China, by taking whatever steps are needed to insure that you receive products that are of uniform good quality. It is important to know what inspection criteria and tests, such as density, microstructure, chemical analysis, thermal expansion, exudation (for fusion cast) should be done, to best evaluate/compare the refractories. Based on my experience of observing the production of AZS blocks in a Chinese plant this year, a variety of observations and comments are presented which document production factors that can cause variations in the quality and integrity of AZS blocks. Although some people believe that sonic testing is not applicable for AZS refractories, the results of the author's limited sonic testing in China indicate that sonic testing is definitely useful in non-destructively evaluating and comparing the integrity of AZS refractories, and with further experience, more definitive analyses should be possible.

In addition to the refractory purchasing opportunities in China, there are significant opportunities for glass refractory research and testing, at several universities and institutes, which may be of interest to foreign companies that have lost staff, budget, and lab capabilities.

REFERENCES

1. Commentator, CCTV-9 (English language channel), Beijing, 1/17/06
2. McDonalds Vice-President, CNBC-TV interview, Oct. 2007.
3. Y.S. Kan (Chinese Businesswoman), Beijing Magazine, April 2008
4. S. Breteau, Asia Inspection Co., USA Today, Money Section, Cover Story, July 3, 2007
5. M. Bellamy, "Global Warning – Finding the Right Supplier in China is not an Easy Task", Door & Windowmaker Magazine, June-July, 2004.
6. S. Baxendale, "Refractories in the Global Glass Market", The Refractories Engr., Sept. 2008, pg. 26-30.
7. G. Evans, "Glass Furnaces: Monitoring Refractory Wear", The Refractories Engr., Nov. 2006, Pg. 7-10.
8. S. Baxendale, "Refractories in a Global Market", Abstract for Soc. of Glass Technology Annual Conf. 2007, Univ. of Darby, UK.
9. Chinese Standard JC/T 493-2001, "Fused Cast AZS Refractories for Glass Furnace".
10. J. Liu and Y. Zhang, "Refractories for Glass Tank Furnaces", The Refractories Engr., Jan. 2001, pg. 21-25

REFRACTORIES FOR A GLOBAL GLASS MARKET

Sarah Baxendale and Nigel Longshaw
CERAM, UK

Recent years have seen significant changes in supply of refractory materials to industries worldwide and an increasing number of products from a variety of new sources have now been released onto the market

In 2007, the world tonnage of refractories was estimated at around 34 million metric tonnes and this figure is expected to rise to around 42 million metric tonnes by 2010 with increased market value and good profit potential[1].

Particularly for the glass industry in the next 2-3 years, Chinese refractories for example, will be sold in India, South East Asia and parts of Europe [2] and European companies now have manufacturing facilities in this part of the world.
As a result of this glass producers have a much wider choice of refractory sources now and competitive pricing makes these new refractories an attractive option. [2]

Many manufacturers are learning to close the quality gap between materials sold on the domestic market and export products[2] and may do so in a relatively short time calling on lessons learnt from other areas of the world as they went through the same process.

However it is still extremely important to carefully assess the widening range of refractory materials for the glass industry to ensure that they give the performances required for cost effective, long term glass production.

This paper looks at some of the methods to assess critical service performance of materials for glass making applications and highlights the potential impact on glass production should refractories not perform to the required standards. Many of the issues highlighted will be familiar to glass makers and whilst it may be felt that these issues have been resolved over the years, with so many new materials available a review at this stage, was thought to be appropriate.

First of all we will, look at the some figures for refractory usage across high temperature industries. Figure 1 shows that the worldwide steel industry still dominates the demand for refractory materials consuming over two thirds of worlds' refractories. In comparison the glass industry consumes a relatively small portion of refractories annually at around half a million tonnes.

As Figure 2 shows this annual consumption of refractories by the glass industry amounts to a total sales value of approximately 0.65 to 0.85 billion dollars of which fusion cast glass contact refractories represent around 45%, followed by basic refractories such as those used in regenerators.

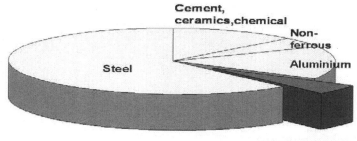

Figure 1 - REFRACTORIES – the glass industry market

TOTAL annual sales
~US$ 0.65-0.85 billions

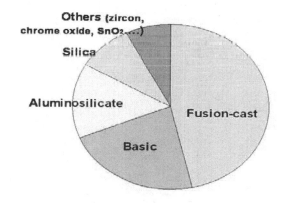

Figure 2 - REFRACTORY SALES TO THE GLASS INDUSTRY

Overall however these sales represent relatively small proportion of the total annual world market value.

Nevertheless glass producers still demand high quality refractories for their furnaces to improve life and hence reduce manufacturing costs.

World glass production in float, container, fibreglass and specialty areas currently stands at around 100million tonnes per annum. (Figure 3)

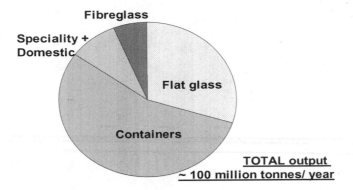

Figure 3 - GLASS INDUSTRY SECTORS

All figures exclude China, however a 2005 estimate, suggested that float glass production stood at around 20M tonnes, 50-60% of the rest of the world.

There are 300 float lines in the world and half of these are in China.

Around 5 kg of refractories are consumed per tonne of glass produced and the cost of these refractories is approximately $8.5 per tonne of glass produced. Cost per tonne of glass is equivalent to 2% per tonne of yield. Refractories are therefore normally the largest single component of furnace build and repair costs.

Their performance can influence furnace output and product quality and hence process economics. The deterioration of the refractories determines the life of the furnace and, as consequence, the lining repair schedule.

These materials are therefore an essential component of the glass making process where they are key contributors to business success.

Figure 4 shows some of the important areas in terms of performance

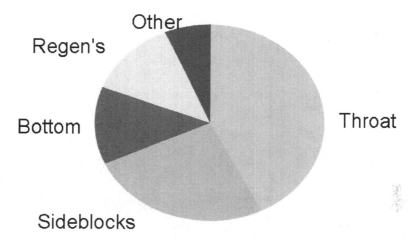

Figure 4 - GLASS FURNACE WEAKNESSES

The glass contact regions namely the side blocks and throat (of container glass furnaces) shown here are the most vulnerable areas of the furnace as glass corrosion is at its most aggressive here.

For most furnaces these critical regions are most likely to be life determining, so it is extremely important to correctly evaluate resistance to high temperature corrosion.

There are numerous tests available for the assessment of refractories in contact with molten glass and these can be divided generically into static or dynamic tests.

In static corrosion tests, the refractory specimen is fixed in position via cement allowing no specimen movement in the molten glass. In the dynamic test there is continual movement of the specimens in the molten glass.

Static tests are useful for comparing the relative performance of glass contact materials however for critical, furnace life limiting applications a 'dynamic test' is preferable.

At ambient temperature the glass cullet is placed in platinum crucible and the furnace is then closed and heated. Once the test temperature is reached, samples are lowered into glass for testing. Assessment of corrosion is by measurement of change in volume.

The dynamic test more accurately reproduces the type of glass line and other high wear corrosion experienced in furnace operation. Corrosion is often ten times higher at a triple interface, such as the glass line, than in other regions of the furnace.

Figure 5 shows the performances of three fusion cast refractories in soda lime glass. As can be seen the dynamic test more accurately reproduces the enhanced corrosion at the glass line and other high wear areas seen in service.

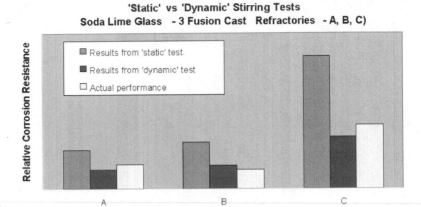

Figure 5 - EVALUATION OF CORROSION RESISTANCE TESTS

The rotation of the samples during the test prevents the formation of a protective boundary layer at the glass / refractory interface, which, in operating furnaces normally only forms in areas of quiescent or stagnant glass flow. The formation such boundary layers in the static corrosion test can restrict its application.

The static test can be used for simple comparisons, but dynamic tests correlate much more closely with in service performance as shown in the graph.

Accurate corrosion assessment is therefore extremely important when selecting new glass contact refractories. What is also important for AZS materials, because of their varying composition within a block is to compare pieces from similar locations within the blocks.

Figure 6 shows the variation of corrosion volume loss with temperature. At 1550°C three AZS materials D, E and F were compared using the dynamic corrosion test. From datasheet values the materials appeared similar however Material F gave a higher than expected volume loss at 52%, the reason for which was then investigated further.

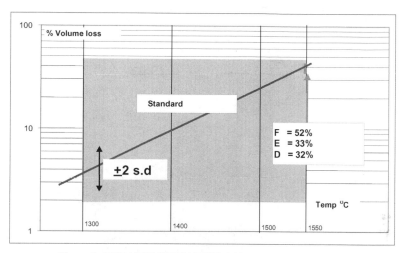

Figure 6 - DYNAMIC TEST,1550°C, 24 hours, AZS, FLOAT GLASS

Further analysis (Figure 7) showed that Material F contained the highest percentage of non AZS components in the glassy phase at around 8.5%, compared to values of just over 7% for material E and just over 6.5% for material D.

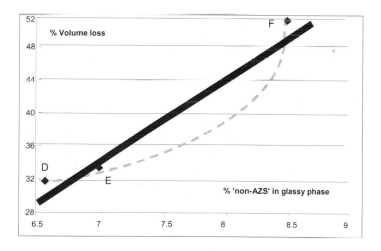

Figure 7 - CORROSION VOLUME LOSS: COMPOSITION OF GLASSY PHASE

The glassy phase is overall the least refractory component of AZS materials and the route through which chemical attack takes place.

As well as considering corrosive effects of glass on the wear of the glass contact refractories we must also consider the potential impact of the corroded refractory on glass quality.

A significant source of glass loss can be attributed to solid, vitreous and bubble faults originating from furnace refractories.

Figure 8 shows the typical data sheet information for a new 33% zirconia AZS fusion cast material that was being considered for use in a container glass furnace.

Chemical Analysis	%
Al_2O_3	49-50
SiO_2	<15
ZrO_2	34-35
Alkalis	<1.5
Physical Properties	
Bulk Density g/cm^3	>3.4
Apparent Porosity %	<2.5

FIGURE 8 - TYPICAL DATA SHEET INFORMATION FOR A NEW 33% ZIRCONIA
AZS FUSION CAST MATERIAL

This data is comparable to other similar products available on the market and from the data its performance would be also expect to be comparable and nothing untoward would be expected.

However when assessed for high temperature glassy phase exudation by firing to 1500C in air and measuring volume changes you get a bit of a surprise (Figure 9)

The material (H) gave higher than expected values for exudation above the acceptable level for this type of material. The exudate is shown on the surface of the test sample.

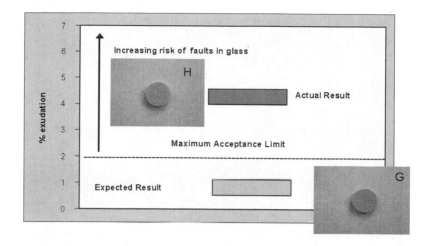

Figure 9 - EXUDATION PERFORMANCE

A result below the maximum acceptance limit of 2 % for superstructure material was expected, but measured results gave values in excess of 4% and above the acceptable limit leading to increasing risk of glass faults.

As is well documented the difference in exudation performance relates to the extent of oxidation of species within the AZS refractory during its manufacture. A poorly oxidised material has the potential to produce high exudation.

Use of this material in the proposed application could have a catastrophic effect on glass quality during the initial stages of furnace operation and increasing the risk of glass defects such as crystalline, dendritic zirconia faults in the finished product.

Assessment of blister formation, derived from impurities such as ferric oxide in AZS materials is also important because of the impact on glass quality.

This involves firing of a refractory disc in contact with glass and assessing by microscopy the number of blisters / cm2 at the glass / refractory interface.

Figure 10 shows two materials, I and J that were assessed.

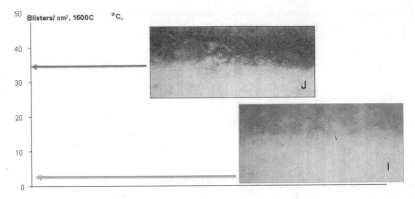

Figure 10 - BLISTER TESTS: FUSION CAST AZS

Material I showed low blister formation. Ideally low single figure values are acceptable. However the high blister count at around 35 /cm^2 would make this material unsuitable for use.

In addition to assessing the performance of glass contact refractories and their effect on glass quality, careful assessment of materials in other regions is essential to ensure smooth, continuous operation.

Figure 11 shows the typical data for silica which was used in the crown of a float furnace.

	Current Silica	New Silica
SiO$_2$%	95.8	96.0
Al$_2$O$_3$ %	0.58	0.54
Fe$_2$O$_3$ %	0.85	0.9
CaO and MgO %	2.25	2.31
Na$_2$O and K$_2$O %	0.20	0.5
Bulk Density, g/cm^3	19.7	20.8
Apparent Porosity, %	1.82	1.78
CCS, MPa	34.1	32.5
PLC @ 1600°C, %	0.3	0.35

Figure 11 - SILICA REFRACTORY FOR FURNACE CROWNS

As we can see the chemical and physical properties of the new material, are generally similar to the currently used silica product. So from the data sheet no surprises are apparent.

However during the heat up of the furnace the crown expansion grossly exceeded expectations making control difficult and resulting in the distortion of the crown and opening up of longitudinal joints along the crown. Potentially these could lead to long term maintenance issues and glass quality if alkali condensation causes rat holing in these joints.

The expansion curves of the two materials in Figure 12 clarify the observations. The curve for the current product is typical of high quality crown silica. However the new silica continued to expand as a result of under firing and therefore high levels of unconverted quartz remained.

So when selecting new products and for quality control purposes it is important to test critical in-service properties. Although silica is relatively cheap compared to the fusion cast glass contact refractories, it is still critical in terms of furnace operation, furnace life and glass quality.

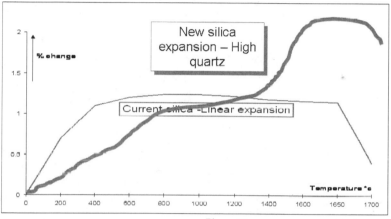

Figure 12

Another critical performance area within float glass furnaces over the years is the tin bath. Today around 95% of the world's flat glass is produced by the float process and this area has its own critical set of performance indictors.

In particular the performance of tin bath bottom blocks is extremely important. Blocks have mainly been based on alumino silicate fireclay.

In particular the performance of tin bath bottom blocks is extremely important. Blocks have mainly been based on alumino silicate fireclay.

A variety of issues have been observed in this area of the float furnace over the years. Nepheline peeling is still the most common problem seen.

Nepheline formation leads to volume expansion and thermal expansion mismatch which leads to detachment of material.

Alkali penetration tests were carried out on three materials over an extended period and following this the degree of penetration is assessed either visually or by microscopy.

Another issue in tin baths is bubble formation on the surface of the glass due to thermal transpiration of blocks when mean pore diameter is smaller than diameter of gas molecules.

This is characterised by the hydrogen diffusivity test and examines the pressure build up in mm H_2O in samples of bath block. A cut off level of 150mm H_2O is used. It is also important to examine different areas of the block as diffusivity may vary.

The corresponding values of materials K, L and M are shown Figure 13.

Material	Alkali Penetration 800°C- 1000°C 7 days	Hydrogen Diffusivity mm H_2O
K	No	57
L	No	142
M	Yes	6

Figure 13 - ALKALI PENETRATION / HYDROGEN DIFFUSIVITY

As with many refractory materials a balance of properties is required to optimise performance.

Figure 14 aims to show the inverse relationship between permeability (i.e. risk of alkali penetration through LARGE pores) and H_2 diffusivity (i.e risk of H_2 transpiration through SMALL pores.) This is the curved line.

The area of 'acceptable' balance between these 2 parameters (which can only be defined by experience) is within the square shown.

Figure 1

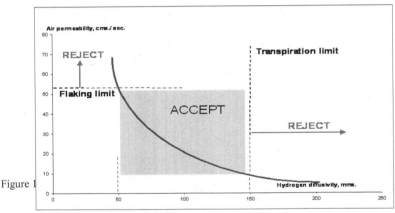

Figure 14 - HYDROGEN DIFFUSIVITY - BLOCK PERMEABILITY

Recent studies[3] have indicated that the amount of glassy phase in the block plays an important role as to whether peeling occurs as a result of nepheline formation.
So assessment of the glassy phase should also perhaps be carried when selecting new bath block refractories.

In addition to assessing the performance of increasing numbers of fireclay bath blocks on the market, a new generation of calcium aluminate blocks is now available on the market, hence more choice still for the glass maker!

So there are many new refractory opportunities for glass producers, but it is critical to assess the in-service performance of the ever increasing choices of materials available on the market as data sheets can only provide a limited amount of information

Independent evaluation of refractory products therefore provides the benefit of improved confidence to the glass manufacturers, refractory producers and furnace engineers alike.

REFERENCES

1. Refractories – A Global Strategic Business Report – June 2007, Global industry analysts, Inc.
2. Burning Issues, Asian Glass, pp55-57, Apr May 2007.
3. Schmalenbach, B., Weichert, T., and Santowski, K., Development of Refractories for the Tin Bath Bottom of Float Glass Lines, RIII Bulletin>3>2006, pp 36-42.

HISTORY OF HOT REPAIR – PAST, PRESENT AND FUTURE

David T. Boothe
Allstates Refractory Contractors

Over the last 40 years with the improvement in refractory, combustion and control technology and better scheduling, the life of an average furnace has increased approx. 2.5 times. Now instead of a scheduled repair every 3 to 5 years we are typically at 8 to 10 years. When you repair every 3 to 5 years, the necessity of midterm repairs was not the focus of a company trying to get the maximum life time out of these units. However, over this same period of time, the need for hot repair services has grown dramatically and the technology of these types of service has advanced as well.

Figure 1 - Crown Repair (hanging brick)

My first exposure to hot repair was the summer of 1969 when Paul Wood the plant manager at Thatcher Glass, Elmira came to a group of us who were working either in the basement as cullet handlers or on the machine floor as sweepers or mold handlers to assist. A portion of a regenerator crown on a large sideport had collapsed and left a gaping hole in what appeared to us to be the hinges of hell itself.

The idea was to hang brick on a grid of steelwork into the hole progressively closing it up one brick at time. The unanimous quotation from us was a resounding "You want to do what!?" However, being young, somewhat fearless and just plain ignorant of the potential hazards, we forged ahead. With great effort and lots of sweat it was accomplished. The next step was something new and untried to my knowledge at the time. The then National Refractory Group came in and pumped a sealcoat over the hung brick giving us a furnace that continued to produce saleable glass for an additional 18 months.

This was the first time that a crown repair of this style was successfully attempted. Being my first exposure to hot repair and being a ceramic engineering student, I was amazed at both the happening and the result. This event was to be the beginning of a career highlighted by doing repairs on operating furnaces that were about to expire and yet through shear imagination and technology were given extended lifetimes.

I will try to take you on a journey through these past decades and give you a tour of things that have been accomplished by folks who said "Let's try this rather than shut it down".

Figure 2 - Crown Repair (pumped sealcoat)

The aforementioned crown repair would over the years come to be common place and even be used to save a float furnace regenerator with a enormous hole some 15 feet across and 18 feet long. Crown overcoats would be used to seal crowns from alkali attack, save crowns whose life was in jeopardy, and eventually will be used to cast large entire crowns in place.

Speaking of regenerators, they have had a multitude of problems over the years especially as we continue to press for more and more tonnage. The most typical of issues is plugging. In years past, many things including dripping water through the crown to disintegrate carry over and refractory have been devised to open or keep open passages for air flow. Amongst the ideas conceived and implemented were the following:

1) Settings were engineered to allow access from the outside during operation for steel rods or compressed air pipes to be introduced to keep the flues open.

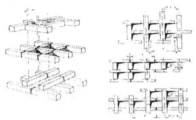

Figure 3 – Checker Plugging (rod or blow out access checker setting)

2) Later in campaigns, exhaust flues were built on the outside of the regenerators to allow gases to circumvent the plugged areas.

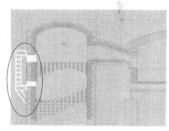

Figure 4 - Checker Plugging (exterior flue)

3) This same concept would be used on the interior, by removing checkers at the target wall and creating an inner chase to get around the plugged area.

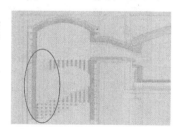

Figure 5 - Checker Plugging (inner chase or flue)

4) Next the removal of entire rows of checkers was done by raking them out through holes in the targetwall eliminating carry over plugged checkers.

Note: All the above will reduce the air preheating potential!

Figure 6 - Checker Plugging (removal of entire row of checkers)

5) Several companies started providing a "blowout" service that attempted to remove or reduce the plugs in the passages through inserting a high pressure compressed air nozzle and blowing up or down the passages. Others provided a "burnout" service which utilized a high pressure gas-air burner to try to melt the plugs out of the affected flues.

Figure 7 - Checker Burnout
(burner installed in access opening)

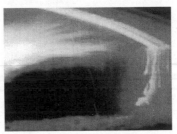

Figure 8 - Checker Burnout
(looking under rider arches during burnout)

Still others in recent years have utilized an extremely high pressure (8000 to 10,000 psi) water blast to wash out the flues. This has limited effect as the water turns to steam rather quickly and is detrimental if the nozzle is not place directly over the flue or moved without reducing the high pressure!

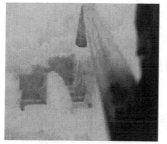
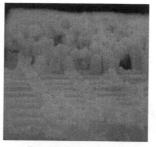

Figure 10 - Checker Washout
(high pressure water lance)

Figure 11 - Checker Washout
(prior to checker washing)

Figure 9 - Checker Washout
(after checker washing)

6) By far the most intrusive, most expensive and most effective of all options for severely plugged checkers is the hot checker change. This entails the complete removal and replacement of a portion or entire checker pack.

To do this one must block off the ports and put the furnace in hot hold by building other temporary exhaust flues and utilizing the main burner system with an outside combustion air source or an outside heat-up service to provide the needed heat.

Figure 12 - Complete Checker Replacement

Access to the checkers is gained by removing the panels at either end of the regenerator or cutting access panels in the targetwalls. The checkers are removed as they would be during a cold repair or by hand, depending on the internal temperature. Once removed, the rider arches are inspected and repaired and new checkers are installed again as they would in a cold repair scenario.

Upon completion of the installation, the regenerator openings are closed and sealed. The ports are slowly opened to let heat gradually enter the checker chambers and a slow heat up schedule will follow, allowing the checkers to come to operating temperature. Once at operating temperature, the ports are resealed and the furnace put back into service. Depending on the size of the furnace this entire event would take ten days to two weeks to complete.

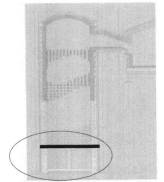

Another issue that has had some interesting solutions is sagging or falling rider arches. This condition has some disastrous side effects! One innovative solution is to cut a slot in both walls next to the damaged arch and insert water coolers to take up the load as the checkers begin to sag. Another solution is to jack the arch up slightly and build a supporting pier(s) under the affected portion of the arch.

Figure 13 - Temporary Checker Support (water cooled)

As glass contact refractory improved; and lifetimes increased, definitive weak spots began to appear. One of these was the throat at the covers and facers. Hence, throat overcoats began to be conceived using a wedge configuration where two or three blocks were slid into place from either side over the existing cover between the facers. This procedure required the throat binding and all cooling to be removed.

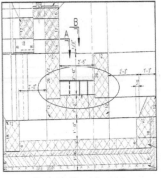

Figure 14 - Throat Overcoat

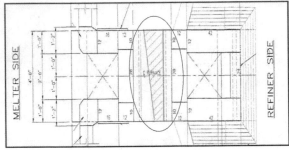

Figure 15 - Throat Overcoat

To accomplish this, the glass had to be frozen on both the melting and refining sides of the throat. Water lances were inserted into the molten glass either through the peep holes or through holes drilled in the crown or breastwall. Water was pumped into these areas freezing the glass. This will remove the head pressure on the blocks and allowed the steel and cooling to be removed safely.

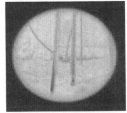

Figure 16 - Freezing Glass (water lances in glass)

Once the wedges were in place, binding was reinstalled and cooling replaced. At this point the water on the lances was turned off and as the glass melted out, the lances were extracted and access

holes closed. This procedure would typically extend the furnace life by 18 to 24 months. Needless to say the money from the sale of glass would more than outweigh the cost of the repair. However, the downside is that the glass coming from the throat will be substantially hotter and may cause forming issues.

Figure 17 - Freezing Glass
(water lances inserted through the crown)

This same procedure would become useful in many other applications that we will cover later in this paper. In addition, the actual removal and replacement of the entire throat can be accomplished in the very same manner. Water lances are inserted on both sides of the throat and the glass is frozen. This procedure freezes the glass within the throat as well. Then binding steel is removed and the entire throat refractory package is demolished. New refractory is installed and the binding is replaced. One small trick is to pack the throat with raw soda ash. This will lower the viscosity of the glass that enters the throat as it heats up and accelerate thawing. In addition, bubblers can be added in front and within the throat to move the molten glass through the throat faster.

While we are on the subject of freezing glass, let's visit the areas that can be accessed without fear of glass leakage through the appropriate use of water. Removal and replacement of sidewall overcoats or even the blocks themselves, addition of forehearth entrances or repair of forehearth superstructure and glass contact, removal and replacement of electrodes and/or holders, removing bubblers and associated blocks are all possible, with most have been completed without incident. As

we can see in the photos an entire doghouse glass contact refractory was removed and replaced while the furnace was held at a reduced temperature full of glass. This entire procedure took only five days glass to glass.

Figure 18 - Doghouse Rebuild

In another case, about fifteen years ago through an error in engineering, a forehearth began to subside shortly after it was filled with glass. It was found that the interface temperatures were too great for the insulation refractory to withstand, they began to deform making the bottom glass contact open up and glass leak into the cavity left by the subsiding insulation brick. It was a new unit and the owners did not want to shut it down and replace it. Through the strategic use of water cooled picks, the temperature at the interfaces was lowered and the forehearth structurally saved and run for 7 years.

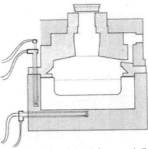

Water jackets and direct misting (sprays) have been used to cool refractory as it becomes thin or at potential glass leak areas. The use of water jackets is preferable to me. However, the direct misting is much more effective in cooling, but significantly more messy and makes the area hazardous and uncomfortable to work in! The use of water jackets has allowed some furnaces to operate for several months with the frozen glass acting as a containment surface.

Figure 19 - Forehearth Repair (water cooled)

For a moment let's discuss water jackets. There has long been a discussion of solid multipass versus open looped water jackets for cooling. In my experience and opinion the solid multipass water jacket is the best. Several reasons come to mind, but the most viable of all is the fact that, if the refractory behind the jacket deteriorates to a point where glass will start to ooze out, a solid water jacket will most likely stop the glass advance. An open loop one will not stop such a leak.

Within the last decade improvement in monolithic refractory has lead to its increased use in hot repair scenarios. It has been used to overcoat the exterior of crowns, exhaust stacks, flues, ports and upper structure walls and at times hot gunned onto the interiors of the same. Some examples of this are:

1. A section of a large float crown was in eminent danger of collapsing. A section 36 feet wide and 20 feet long had been subsiding for several weeks. No other solutions had been conceived or used to overt the potential disaster and allow the furnace to continue to operate. After careful consideration and lots of planning, a self supporting crown overcoat of fused silica was installed. This would provide a secondary crown to cover the opening that the collapsing crown would leave and hopefully stave off the impending collapse for some period of time.

Figure 20 - Crown Overcoat (shotcreting)
courtesy of Magneco Metrel

The particulars are too involved to go into here, but suffice it to say that the furnace continued operating for another 7 plus years before it was taken down for repair.

2. Another instance where monolithic materials saved a furnace from potential disaster was an endport container tank whose backwall was about ready to fall in. It was hot repaired by casting a new backwall over the existing one. A feat that to my knowledge had never been attempted before and was successfully implemented by some very progressive and innovative people in the company's engineering staff. Typically, this wall would have shut the furnace down. However, the entire wall was formed and poured in sections from the tuckstones to the main crown encapsulating the existing support steel and port entrances.

The existing buckstays and binding were used as forming supports and new steel was erected behind the existing. The castable was pumped in sections roughly 18 inches high and allowed to set before the next section was added. This reduced the head pressure on the existing wall and allowed the new wall to carry its weight without sagging or buckling.

Figure 21 - Stack Overcoat (cast in place hot)

The port entrances melter side and the existing binding were totally encapsulated. The small opening at the main crown was enclosed using hanging brick and overcoat. The existing backwall fell in, but the furnace continued to operate for several months. This same procedure has been utilized in breastwalls, frontwalls, targetwalls, flues, and exhaust stacks, etc. with great success. Hot gunning of monolithic materials has also been used to overcoat the interiors of mainly exhaust flues and stacks and at times regenerator walls to fix

Figure 22 - Stack Overcoat (cast in place hot)

worn spots or replenish entire thin wall sections. This process is faster, but has a limited life expectancy.

An interesting sideline to this type of repair is the rebuilding of port floors and sills. This repair is usually done in conjunction with the repair or replacement of severely damaged or loss of the port sill blocks. The process is similar to the hot gunning technology without the use of high pressure material spraying. The port floor is raked out as well as it can be. Sidewall overcoat blocks are installed under the ports and elevated past the existing port sill to create a small dam at the port snout. A castable material, usually with some addition of a catalyst, is pumped carefully through a water cooled pipe and like concrete is leveled at the desired height. Working times here are critical as is the choice of material. However, the success with this type of repair is very good and extends port life and protects the port support steel! In addition, this repair method has been recently adapted to melter floor repairs. The glass is drained out and the furnace cooled to a hot hold temperature. Self leveling castable is pumped onto the surface and allowed to set. Any holes are filled or imperfections are coated over and a new surface of the desired thickness is installed. As with any repair of this type in a glass containment area, there are glass defects that will follow the repair, but these are generally short lived and do not limit the productivity of the unit once they subside.

In a discussion on the subject of fixing holes, walls, etc., let us not forget the ceramic welding technology that was developed in Europe and brought to the U.S. Simplistically, ceramic welding entails the use of water cooled pipes that are inserted into the hot space and directed at the area in need of repair. A combination of refractory material and a temperature ignited metal are pumped through at high pressure and expelled just inches away from the surface to be repaired. The metal burns at a temperature that melts the refractory powder and sprays the molten material onto the affected area. At once it adheres, cools and coats the surface.

Figure 23 - Ceramic Welding
courtesy of Fuse Tech, Inc.

Several passes are needed to build up the coating thickness and fill the hole. Welding like hot gunning has a limited life and requires reapplication several times over the course of months to extend the campaign.

Figure 24 - Collapsed Arch (before and after)

At last we come to the part of hot repair that is near and dear to my heart. This is the direct replacement of refractory in a severely worn or open area of the furnace. One of the most interesting and innovative advances in hot repair technology in the last 20 years is the use of the hydraulic-driven, water-cooled diamond-segmented chainsaw. This tool has revolutionized the removal and replacement of worn or damaged refractory and the process of cutting a hole in any refractory surface. This piece of equipment utilized by the concrete industry for decades was appropriately adapted for use in the hot repair industry by the engineers at Anchor Glass Container Corp. and advanced by the people at D. Boothe & Co., Inc. (The predecessor to Allstates Refractory Contractors, LLC.) This technology has grown in usage and value significantly over the last ten years.

Figure 25 - The "Saw"
(cutting a slot for water coolers)

The Saw as it is affectionately called, allows a hole in any refractory to be cut in a square or rectangular shape. This is an obvious advantage since the material used to fill it comes in rectangular shapes, thus allowing the hole to be closed with a tight joint and level support. The nuance though is the correct

Figure 26 - The "Saw"
(cutting a seam in the wall)

use of the device to limit the contamination of the glass when cutting holes above glass containment areas. This one piece of equipment has reduced hot repair times, limited exposure of the people doing the repair, and allowed a better, much more efficient use of repair materials!

The direct replacement of refractory in walls (including sidewalls), crowns, and ports, in both the regenerator and melter is the best way to ensure longer furnace life. The refractory is most like the original and will in most cases, with proper choice and installation give the repaired area renewed structural strength, allow insulation to be used and provide longer life than most other repair methods. It can be accomplished in segments and usually eliminates the need to hamper production. These repairs may be followed by welding the interior and/or coating the outside with a monolithic or toweled on material to completely seal the joint structure inside and out.

Figure 27 - The "Saw" (replacement of brick)

There are other projects and methods that could be described here, but time will not allow that. Suffice it to say, the future of hot repairs is bright and the technology of materials and their application will continue to be a priority for research and development. "Star Flo" is just one example of this. The things that will be done will be the result of imagination; engineering, technology and a willingness of management to say let's try something other than a shutdown. Recently, we developed and implemented a new technology for the application of bonded AZS patch material to fill horizontal crevices in crowns and vertical walls that are not open to the outside. A type of air driven bazooka is loaded with patch material in dough like consistency and shot into the opening filling the void and closing up the hole. An unconventional approach, but highly effective and better than welding for filling the gap.

With the constant evolution of furnace designs, new problems will arise and it will be the hot repair companies charge to conceive a method to repair them mid-term. Nothing is impossible; probability is bounded by risk potential and the experience and imagination of the people willing to make it happen. As in all of these technologies, great care, good planning, safety, and experience are the keys to success! The contractors and in house staff must be constantly aware of the dangers of working in these environments and never be hasty in the application of their trade. All must be vigilant, well trained, have proper PPE and knowledgeable about the materials, equipment, and procedures utilized in the application of the technologies described here. In addition, the companies that hire the people who perform these kinds of miracles need to be diligent in their understanding of the need, the time required, potential risk, cost, and due diligence in selecting those who apply these technologies. More than one has been disappointed by not doing the very things we have discussed here.

In having said that, there are no crystal balls or magic wands to fix some problems. Things happen that are out of the control of the trades people, engineers and operators. Be cognizant that early detection and proper repair, as the doctor says, is the best chance for long life and disaster avoidance!

WATER-COOLED REFRACTORY SHAPES FOR HARSH FOREHEARTH SUPERSTRUCTURE APPLICATIONS

Walter L. Evans, II, Luke Evans, and Christopher W. Hughes
Special Shapes Refractory Company

ABSTRACT
Glass furnace forehearths superstructures can be environments of rather harsh conditions. This paper will serve to illustrate that water-cooled refractory shapes are a viable and cost-effective solution for previously unmanageable problem areas requiring innovative refractory techniques. This approach is a joint effort between Special Shapes Refractory
Company and an end-user to resist the deleterious effects of Sodium Borate attack, thermal cycling, hazardous environments, and loss of production.

INTRODUCTION
A client contacted Special Shapes Refractory Company and related they were having a problem with a forehearth superstructure pour point hole, where borosilicate glass is introduced. This area is prone to several problems, including :

- Vapor attack from Sodium Borate glass
- Condensate
- Accumulation of slag and re-melt
- Damage to adjacent refractory due to thermal cycling

Figure 1   Example of forehearth pour point prior to use of water-cooled blocks

These problems resulted in the following issues :

- Unscheduled downtime
- Loss of production
- Increased material and labor costs of hot repairs

PREVIOUS SOLUTIONS
Previously, a variety of refractory oxides had been employed to combat the deleterious effects at the pour point. These included :

- Bonded AZS GC
- Zircon GC
- 30% Chromic Oxide
- 50% Chromic Oxide
- Zircon GC with a water-cooled insert
- Alumina-mag spinel

We were advised that all of these products offered some level of success, with a 12" thick Zircon GC block with a water-cooled insert offering the longest life at approximately 21 months before a hot repair was required. However, the client was unhappy with the usage life and searched for an improvement.

INNOVATIVE SOLUTIONS
After several discussions together, SSRC and the client arrived at an idea for a water-cooled refractory shape that could resist the chemical and thermal attack mechanisms that had destroyed all previous pour point refractories to date. A pipe layout was designed, a refractory oxide was selected (35% ZrO2, bonded AZS), and metal fiber reinforcement was employed for the shape. The name chosen was the B² Water Cooled Refractory.

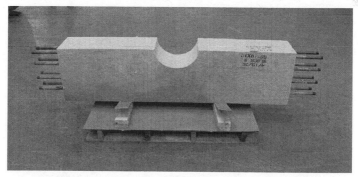

Figure 2   Water-cooled refractory shape prior to installation

Figure 3   Water cooled refractory block in service

SUCCESS
In July, 2007, after 18 months of service, the test block was removed for inspection. The refractory exhibited only 1 to 2 inches of wear.

Figure 4   Inspection of shape after 18 months service shows minimal wear

Reasons for the success can be attributed to the following:

- The uniform cooling effect minimizes the effects of the Sodium Borate vapor, condensate, and slag penetration. Attack is limited to the surface of the refractory blocks, thereby protecting the internal structure.
- The addition of special stainless-steel needles reinforce that create improved flexural strength and maintain a connected body when exposed to thermal cycling. This also reduces "peeling" and "slabbing" in the reacted and densified hot face refractory.

CONCLUSIONS

With the advent of the water-cooled refractory block, pour point life is now expected to match the life of the forehearth. With this being the case, new areas are presenting themselves as candidates for this technology in order to meet cost and production goals.

Pipe layout and alloy can be modified to adapt to a wide variety of shapes, sizes, and applications. The water-cooled concept is expected to offer an innovative way to balance all zones of the forehearth to reach a full campaign.

A NEW SILICA BRICK WITHOUT LIME BONDING

Götz Heilemann[1,a], Dr. Stefan Postrach[1,b], Dr. Rongxing Bei[1,c], Dipl.-Ing. Bernhard Schmalenbach[1,d] and Dipl.-Ing. Klaus Santowski[2,e]

[1]RHI Glas GmbH, Abraham-Lincoln-Str. 1, D-65189 Wiesbaden, Germany

[2]RHI AG Refractories Technology Center, Magnesitstr. 2, A-8700 Leoben, Austria

[a]goetz.heilemann@rhi-ag.com, [b]stefan.postrach@rhi-ag.com, [c]rongxing.bei@rhi-ag.com, [d]bernhard.schmalenbach@rhi-ag.com, [e]klaus.santowski@rhi-ag.com

ABSTRACT

STELLA GNL -a new silica grade- is characterized by a SiO_2 content of 99 wt.% and has no lime in the bonding phase. In contrast to this, the standard silica grades exhibit a SiO_2 content between 95 and 97 wt.% and a lime content of approx. 2.5 wt.%. Based on the improved chemical composition, the new Silica grade has a higher corrosion resistance against alkalis and is a suitable grade for the crown of oxy-fuel fired soda-lime furnaces. For conventional firing (air-fuel), this new grade can be used at higher temperatures than standard silica grades.

INTRODUCTION

With the development of oxy-fuel firing, the limits of traditional silica materials concerning their corrosion resistance became evident. Shortly after the first installation of silica crowns in oxy-fuel fired furnaces intensive investigations regarding the reasons for the premature wear started. During the investigations of this phenomenon it was demonstrated that the alkali concentration -especially the NaOH concentration- in oxy-fuel-fired furnaces is 1.5 to 6 times higher compared to traditional regeneratively fired furnaces [1]. The reason for this is the absence of nitrogen in the furnace atmosphere whereby the amount of exhaust gas is reduced. Additionally, the higher vapour content in the combustion space causes increased NaOH evaporation.

Besides the new altered furnace conditions, the material characteristics of standard silica grades are important to understand this problem. The typical chemical composition and physical characteristics of silica bricks are detailed in Table 1. The chemistry is characterized by a SiO_2 content between 95 and 97 wt.% and a CaO content of approx. 2.5 wt.%. Together with SiO_2, CaO forms the bonding phase wollastonite ($CaO*SiO_2$). Furthermore, thermal expansion is an important characteristic. This is mainly influenced by the trydimite and cristobalite phases, which lead to a high expansion in the temperature range from 100 °C up to 400 °C. If the residual quartz content is too high an additional expansion is possible during the heat-up procedure of a furnace at high temperatures.

Table 1: Typical characteristics of standard silica grades

Chemical composition [wt. %]				
SiO$_2$	Al$_2$O$_3$	CaO	Fe$_2$O$_3$	Alkali oxids
95–97	0.2–0.7	~ 2.5	0.3–0.7	0.1–0.3
Physical characteristics				
Density [g/cm^3]	Open porosity [vol.%]	Cold crushing strength [MPa]	Max. thermal expansion [%]	Refractoriness under load [T$_{0.5}$]
1.81–1.87	19–22	30–50	1.3–1.5	1650 °C

Based on the fact of premature wear of standard silica grades under oxy-fuel conditions two developments started. Firstly, the potential of alternative crown materials were investigated, whereby fused cast AZS and alpha-beta alumina were identified as the most suitable solutions. The latter material has also been proven effective under the aggressive conditions in special glass furnaces [2]. Secondly, silica grades were investigated in more detail to fully understand the issue of premature wear. Further investigations have shown that the observed corrosion is influenced by the wollastonite bonding phase [3]. This bonding phase is easily corroded by alkalis, for example NaOH. As a result of the reaction between wollastonite and NaOH, the formation of glassy phase in the silica brick is initiated, which itself starts to dissolve the coarse silica grains. Within a short time the silica brick is enriched with glassy phase and it starts to drip. Subsequently, these reactions have been confirmed by Prof. Beerkens and thermochemical investigations have shown that higher CaO contents in silica grades are critical [4].

Besides these findings, additional practical experience has clearly demonstrated that CaO free silica bricks have an enhanced corrosion resistance: Silica bricks based on fused silica are often used as hot repair materials and exhibit outstanding performance [5,6]. However, the disadvantage of this type of refractory is the shrinkage at T > 1100 °C (i.e. > 1%) [7]. Due to this behaviour, application of this type of material is limited.

Advantages of the new development

Based on the aforementioned experiences, RHI developed a completely new silica brick, termed STELLA GNL, with a SiO$_2$ content of > 99 wt.%. The central aspect of this new grade is a special bonding phase that replaces the traditional wollastonite bonding phase. This specific bonding phase is formed as a result of special manufacturing conditions during the batch preparation and firing process. This bonding phase leads to a silica refractory with outstanding characteristics (summarized in Table 2):

Chemical Composition. The 99 wt.% silica content of the newly developed grade is significantly higher compared to standard silica grades. In addition, due to the fact it is based on synthetic raw materials, the alkali content and Fe_2O_3 is relatively low. However, most importantly no CaO addition is necessary, which avoids the weak wollastonite bonding phase. Laboratory investigations and a practical test in an oxy-fuel furnace have shown that the corrosion resistance is significantly improved compared to standard silica grades (Fig. 1).

Fig. 1: Standard silica grade and STELLA GNL after 3 months in an oxy-fuel furnace

Thermal Expansion Behaviour. The expansion behaviour of the no lime Silica compared to standard silica grades is completely different (Fig.2). Up to 700 °C the maximum expansion is only 0.65%. Furthermore, at higher temperatures the expansion is not significant. Thus the heat-up procedure known to be critical for silica bricks is simplified. Additionally, the typical shrinkage of fused silica grades is not observed.

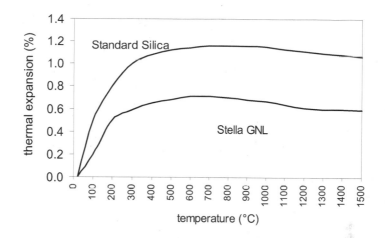

Fig. 2: Thermal expansion of a standard silica grade and STELLA GNL

Creep and Refractoriness Under Load. Due to the reduced alkali oxide content in the new grade, the creep and refractoriness under load hot properties are significantly improved compared to standard silica grades. While standard silica material shows a typical creep at 1600 °C of -0.6% (0.2 MPa/5–25 h) the value for it at 1650 °C is 0%.

Table 2: Typical characteristics of CaO-free STELLA GNL

Chemical composition [wt. %]				
SiO_2	Al_2O_3	CaO	Fe_2O_3	Alkali oxides
99.0	0.1	< 0.05	0.1	< 0.1
Physical characteristics				
Density [g/cm^3]	Open porosity [vol.%]	Cold crushing strength [MPa]	Max. thermal expansion [%]	Refractoriness under load [$T_{0.5}$]
1.83	19.5	35	0.65	1690 °C

Since 2005 RHI has already sold more than 400 tonnes of STELLA GNL for the installation in 10 different glass furnaces. The applications include highly stressed sections of float furnaces and the crowns of high alkali glass furnaces.

REFERENCES
[1] Liang, X., Headrick, W. L., Dharani, L. R.: Current understanding of oxy-fuel glasstank crowns, Refractories Applications and News, Vol. 9, No. 2, 2004.
[2] Gupta, A., Selkregg, K.R.: Performance evaluation of fusion cast alpha beta alumina crowns from industrial oxyfuel furnaces, International Glass Review, issue 2, 2002.
[3] Faber, A.J., Verheijen, O.S.: Refractory corrosion under oxy-fuel firing conditions, Ceram. Eng,. Sci. Proc., 18 [1] 1997.
[4] RHI: Neuer keramisch gebundener Werkstoff für den Oberbau von oxy fuel beheizten Glaswannen auf SiO$_2$-Basis, presentation at HVG-FA2 -Meeting on 22th March 2003 in Niederdollendorf, Germany.
[5] Shamp, D.: In-situ testing of superstructure refractories, Ceram. Eng,. Sci. Proc., 18 [1] 1997.
[6] Kobayashi, H., Wu, K.T., Tuson, G.B., Dumoulin, F., Kiewall, H.P.: TCF Technology for Oxy-Fuel Glassmelting, American Ceramic Society Bulletin, Vol. 84, No. 2, 2005.
[7] Postrach, S., Borens, M.: Characteristics and behavior of the slip-casted material Quarzal, or Is slip-casted fused silica still up-to-date for glass production?, presentation at 48. Internationales Feuerfest-Kollquium, Aachen, 2005.

Combustion and
Energy Savings

AN IMPROVED SOLUTION FOR OXY-FUEL FIRED GLASS MELTING FURNACES

Matthias Lindig
Nikolaus Sorg GmbH & Co. KG,
Lohr am Main, Germany

SUMMARY

Oxygen fuel firing has become one of the competitive solutions for glass melting in the past decades. Environmental pressures, investment costs and individual energy cost conditions are the driving forces for conversion. Numerous improvements in oxygen firing technology have been made in the past, but one of the remaining concerns regarding oxy-fuel firing was the very high flue gas exit temperature.

SORG® has carried out an extensive modeling study to address this issue, which has led to a new furnace design based on known and proven SORG® furnace design features. The SORG® LoNOx® Melter was used as a reference for the development.

As a result of the study an oxy-fuel fired furnace with one radiation wall was developed. The tank utilizes the design features known from the LoNOx® Melter in principle. The dimensions in the combustion chamber and melter tank were optimized during the modeling study.

The furnace design provides a reduction in energy consumption by about 10 %, equivalent to a flue gas temperature reduction of more than 200 °C. Investment cost increase as a result of a footprint increase of about 13 % and the radiation wall will be compensated by energy savings within about 2 years at current energy costs. Additional heat recovery facilities behind the furnace are possible.

1. GLASS MELTING EFFICIENCY – GENERAL REMARKS

The specific energy consumption of conventional fired glass furnaces has been decreased over the last 50 years by at least 50 % with equivalent reduction in CO_2- emissions (Fig.1).

Fig.1: Improvements in energy efficiency and emissions reduction

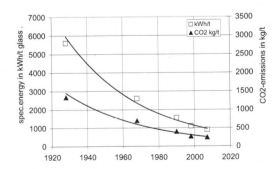

This has been achieved by various improvements in combustion air preheating, insulation and combustion, addressing the heat transfer to the glass bath, and also by increasing the melting efficiency [1,2]. The efficiency of the melting process was raised by tank design modifications relating to the melting, mixing and refining conditions in the furnace. Bubbling systems and all types of booster or barriers are the typical tools for improving the melting efficiency. Examples for melting efficiency improvements based on boosting application are given in the following table (table1).

Endport furnaces
Natural gas fired
50% cullets Emerald green

melting area m²	pull t/d	spec.pull t/m²d	total energy kWh/kg	MJ/t	Boosting kW	%	fossile fuel kWh/kg	total cost Cent/kg
108	246	2,28	1,14	4101	0	0	1,14	2,63
97	296	3,05	1,13	4042	713	5,1	1,07	2,80
108	295	2,73	1,02	3678	1150	9,1	0,93	2,70
100	340	3,40	0,97	3469	1500	11,0	0,86	2,61

| Natural gas | 0,023 €/kWh |
| el.power | 0,06 €/kWh |

Table 1: Container glass furnaces melting efficiency with electric boosting

Compared to the past further improvements in energy efficiency will be more complex, expensive and difficult to achieve.

Benchmark studies [3] and examples of SORG® state-of-the-art in efficiency are in good agreement. Figure 2 shows an example of specific energy consumption versus specific pull. At the moment a specific pull rate of about 4 mt/m².d seems to be an upper limit for the conventional melting process achieved with significant share of electric boosting, with an equivalent energy consumption of below 3500 MJ/mt of glass (50 % cullet, green glass).

A specific energy consumption of 3500 MJ/mt might be close to the physical limit for a conventionally fired furnace with air preheat and without any further waste gas heat recovery.

Despite these improvements, the pressure to progress even further is unrelenting. The cost of fuel has increased dramatically during the last 5 years, and environmental emission limits will result in significant pressure for change in the glass melting process in the near future, with lower emission levels of NO_x and CO_2 necessary.

Today the NO_x emission performance of an improved conventional gas-fired glass melting furnace will be about 800 mg/sm³ on average with somewhat lower values possible for short periods. Lower NO_x emission limits are under discussion in many countries for the future. Conventionally fired furnaces will not be able to achieve significantly lower NO_x emission values with primary measures alone. Additional investment in equipment such as catalytic reactors for DeNOx will be necessary.

2. OXYGEN FIRED GLASS FURNACES – A PRIMARY MEASURE AGAINST NO$_X$

Oxy-fuel firing has been known as an efficient primary measure against NO$_x$ emissions for at least 15 years. In a comprehensive paper [4] Beerkens has evaluated the development and the state-of-the-art of the oxy-fuel firing technology during the past few years. Similar conclusions are also given by Kobayashi [5].

The well-known teething problems were the high peak flame temperatures and velocities, positioning of the burners, high crown refractory wear and increased scum layer on the glass bath. In the meantime burner suppliers have achieved significant improvements, the burner location was reconsidered and the combustion chamber design and refractory assembly was improved. As a result oxy-fuel firing of glass melting furnaces has become a reliable technology with numerous applications in the field of special glasses, fiber glass and container glass, with some examples in the float glass industry as well.

The conversion from conventional combustion to oxy-fuel firing was and is mainly driven by environmental issues, together with consideration of the overall costs situation comprising initial investment costs and operating costs.

The overview given by Beerkens [4] ends with the remark, that the high flue gas temperatures might be one of the key issues for possible future improvement of oxy-fuel fired glass melting furnaces. This thought exactly matches the motivation of the SORG® furnace design group looking for further improvements regarding that type of furnace heating.

Even though the oxy-fuel fired furnace investment might have some advantages compared to regenerative furnaces, the generation of oxygen is still an important factor in any comparison of operating costs. The SORG® intention was to achieve reductions in energy consumption when considering the total furnace heat balance.

(Efficiency improvements should be achieved with a minimum of additional investment cost and no additional heat recovery processing. Any further additional heat recovery facilities located after the optimized process should be an option.

Heat recovery devices such as batch or cullet preheating have been used on oxy-fuel fired furnaces, but further improvements will be necessary due to the special flue gas conditions from this type of furnace. The SORG® approach to this investigation and the final results achieved are presented below.

3. OXYGEN FURNACE COMBUSTION CHAMBER MODIFICATIONS - MODELING STUDY

The flue gas temperature of an oxy-fuel fired melter is significant higher than that of a regenerative furnace after the regenerator chamber. However the heat content of the flue gases is comparable in both cases due to much lower flue gas volume of the oxy-fuel unit see table 2).

Table 2:   Container glass furnaces with different heating systems; comparison of the flue gas heat content

		endport	unit melter	Oygen-gas
pull	t/d	180	180	180
glass			flint	
melting area	m^2	72	113	90
Nat.gas	Nm^3/h	930	1125	921
comb.air/ Oxygen	Nm^3/h	9800	12400	2100
el.boosting	kW	750	750	670
spec.energy cons.	kWh/kg	1,33	1,60	1,34
flue gas temperature	T/°C	520	750	1200
flue gas volume	Nm^3/h	11900	14800	4050
Q $_{flue\ gas}$	kW	2157	4191	1909
Q $_{flue\ gas}$/Q $_{total}$	%	23,2	37,2	20,7

The SORG® approach was to combine the experience with the SORG® LoNOx® Melter and oxy-fuel firing technology. Figure 2 shows a schematic diagram of the LoNOx® Melter, a modified recuperative melter. The key features of that furnace design are the two radiation walls that are used to divide the combustion chamber into two equal sized sections, one heated and a second without combustion, only ventilated by the flue gases escaping at the end of that section.

The second section acts as a kind of internal batch preheating area. Heat transfer is optimized and the flue gases exit with a temperature of about 1200 °C, about 200 °C below the flue temperature of a conventional unit melter. With the LoNOx® Melter it is possible to achieve very low NO$_x$ emission values. A cullet preheating system attached to the melter results in very competitive specific energy consumption compared to regenerative furnaces. A comprehensive summary of that furnace design and performance is given by Sims[6].

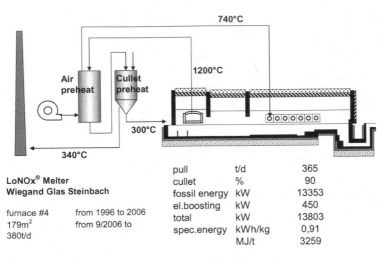

LoNOx® Melter	pull	t/d	365
Wiegand Glas Steinbach	cullet	%	90
	fossil energy	kW	13353
furnace #4 from 1996 to 2006	el.boosting	kW	450
179m² from 9/2006 to	total	kW	13803
380t/d	spec.energy	kWh/kg	0,91
		MJ/t	3259

Fig. 2: Schematic view of the SORG® LoNOx® Melter

The improved oxy-fuel fired furnace design is based on the LoNOx® Melter technology, taking into account that significant efficiency improvements should be also achievable without cullet preheating. The investigation of the new furnace design was carried out on the basis of a mathematical modeling study.

As a first step the reference calculations were carried out for the LoNOx® Melter and for an existing SORG® oxy-fuel fired furnace. The process parameters and furnace dimensions of both units are given in the following table.

Table 3:   LoNOx® Melter and oxy-furnace dimensions and operation data

		Oxy-Fuel	LoNOx® Melter
pull	t/d	400	340
glass		green	green
melting area	m²	153	182
melter length	m	18,3	21,7
length charge end	m	1,4	1
melter width	m	8,4	8,4
spec.pull	t/m²d	2,6	1,86
cullet	%	65	85
cul./batch preheat	T/°C	300	270
boosting	kW	0	500
energy total	MJ/t	3447	3400
Q preheat	kW	0	878
Q total energy	kW	16000	13415

The results of the two modeling tests are given in figure 3.

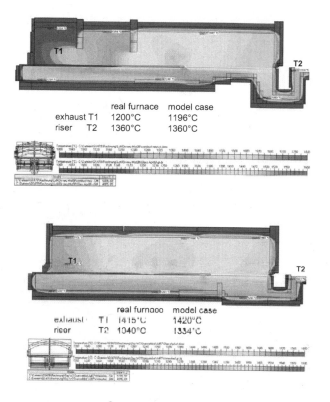

		real furnace	model case
exhaust	T1	1200°C	1196°C
riser	T2	1360°C	1360°C

		real furnace	model case
exhaust	T1	1415°C	1420°C
riser	T2	1340°C	1334°C

Fig. 3: LoNOx® Melter and oxy-fuel fired furnace modeling results

The modeling results were in good agreement with the real furnace operating data.

Based on these results the oxy-fuel fired furnace model was equipped at first with two and then in the further calculations with one radiation wall, whilst dimensions and operating parameters were kept constant. It quickly became obvious that the temperature regime in the charging end could not be held at an acceptable level with two radiation walls. The use of extensive electric boosting to compensate for the low temperature level in the charging end was not considered to be an option as the use of secondary energy sources was to be kept to an absolute minimum.

Several modifications were made to the crown height, the location of the radiation wall and the tank geometry involving the special design features refining shelf and Deep Refiner®.

The final design proposal for the improved oxy-fuel fired furnace in comparison to a conventional oxy-fuel fired furnace is shown in figure 4.

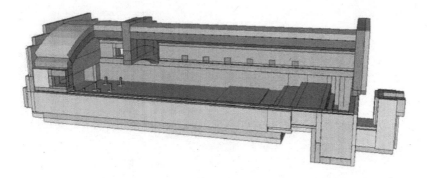

Fig. 4: 3D view of the final design proposal

4. MODELING TEST RESULTS AND PERFORMANCE COMPARISON

The results of the model study are given in figure 5.

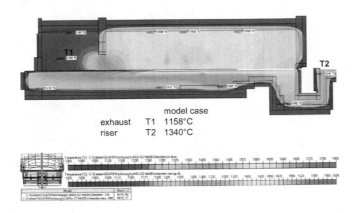

		model case
exhaust	T1	1158°C
riser	T2	1340°C

Fig. 5: Longitudinal section of the SORG® oxy-fuel furnace design proposal

In the calculation of the final design proposal the temperature regime in the charging end was very similar to the conditions we already know from the existing SORG® LoNOx® Melters. The waste gas temperature was even lower at about 1160 °C. At 50 % cullet about 400 kW were used in charging end the test calculation, but it is expected that the charging end boosting system will not be required at higher cullet levels. The reduction in waste gas temperature is equivalent to about 10 % energy reduction.

The special tank design with refining shelf and Deep Refiner® may be necessary in order to ensure good refining. Calculations with standard tank designs have indicated lower melting and fining index numbers.

In comparison to a standard oxy-fuel furnace the investment costs will rise slightly due to the additional need for fused cast material for the radiation wall, and the footprint of the furnace which will be increased by slightly less than 13 %. Key figures from the energy balance comparison between standard oxy-fuel firing and the SORG® modified furnace design are given in table 4.

Table 4: Key figures of the energy balance comparison

		Oxy fired	Sorg Oxy modified
melting area	m^2	137	155
pull	t/d	350	350
cullet addition	%	50	50
spec.energy consumption	MJ/t	3912	3549
	kW/kg	1,09	0,99
spec.melting capacity	t/m^2d	2,55	2,25
Q_{Gas}	kW	15305	13914
Q_{Boost}	kW	500	500
$Q_{flue\ gas}$	kW	-5056	-3581
$T_{flue\ gas}$	°C	1450	1150
T_{glass}	°C	1340	1340

5. ECONOMICAL ASPECTS AND COMPARISON

The additional investment costs for the improved furnace design were calculated and compared wit the annual savings due to energy reduction. The calculation came up with a return of investment after about 2 years based on the today's average energy costs. The competitiveness of the improved oxy-fuel furnace design, and of oxy-fuel firing in general, will depend strongly on the energy cost development in the future. The cost of electrical energy has not always followed the cost development of fossil fuels in the past. From this point of view oxy-fuel firing with an improved furnace design could become more reasonable.

6. CONCLUSIONS

SORG® has carried out a modeling study into the modification of an oxy-fuel fired melter to improve energy consumption. The flue gas temperature level at the combustion chamber exit was the main issue of the study. The known SORG® LoNOx® Melter was used as a reference for the study, with the well-known design features of that furnace design being modified and adjusted to fit the requirements of the oxy-fuel firing. The study produced a solution involving one radiation wall together with the proven design features of the LoNOx® Melter in the tank. The flue gas exit temperature was reduced by over 200 °C, which is equivalent to energy savings of about 10 %. Further waste gas heat recovery measures are optional. Increased investment costs will be compensated by energy savings within a short period of time.

7. REFERENCES

(1)   Evans, G.; Energy efficiency - what next? Glass 10 (2006), 8-11
(2)   Sims, R.: The future of glass melting technology? Glass Worldwide 13 (2007), 30-35
(3)   Beerkens, R.G.C.; Limpt van H.A.C.; Jacobs, G.: Energy efficiency benchmarking of glass furnaces. Glass.Sci.Technol. 77 (2004) 2,pp.
(4)   Beerkens, R:G.C.: Oxygen-fired glass furnaces: expectations and experiences. Glass Machinery and plants 3 (2006), 94-102
(5)   Kobayashi, H; Wu, K.T.; Tuson, G.B.; Dumoulin, F., Kiewall, H.P.:TCF technology for Oxy-fuel glass melting. Am Ceram Soc. 84 (2005) 2, 14-19
(6)   Sims, R.; Pieper H.: The LoNOx® Melter has proven to be a very good choice. Intern. Glass Journal (1996) 88, 49-51

BATCH PREHEATING ON CONTAINER GLASS FURNACES

Hansjürgen Barklage-Hilgefort
Ardagh Glass Group PLC
Nienburg, Germany

ABSTRACT

This paper describes the practical experience with a batch preheating system which was developed in the early eighties of the last century. The system is working with direct contact between the flue gas and the raw materials. Because it is laid out for the operation with high cullet contents it is suitable mainly for container glass furnaces.

Because the emission limits mainly for dust have to be taken into account, the additional installation of a dust filter is obligatory. The impact of the batch preheating on the energy consumption and the pull rate of the furnace is discussed. Due to the direct contact between the flue gas and the raw materials the concentration of SO_x, Chlorine and Fluorine in the flue gas is reduced already in the batch preheater.

The batch preheater has been integrated in the flue gas system of four regenerative furnaces; one system was installed in combination with an oxy fuel furnace. The first installation was started in 1987. Three more followed in 1991 to 1995. Three of the systems are operated still today.

As in any glass melting furnace the energy consumption strongly depends from the cullet content and the pull rate of the furnace practical figures generally an energy saving of 12 to 18 % compared to furnaces without a batch preheater. The pull rate of the furnace can be increased in the same range as the energy consumption of the furnace decreases.

ENERGY EFFICIENCY IN THE GLASS INDUSTRY

In the early 80s there were strong activities in the glass industry to find new possibilities to save energy in the glass melting process. The impact of additional insulation of the furnaces or of the increase of the size of the regenerator on the energy consumption is limited. Additional insulation of the furnace makes not only the construction of the furnace more completed, but has also some negative impacts on the furnace operation and safety, because the observation and control of critical parts are more complicated. The efficiency improvement of any air preheater is not a linear, but an asymptotic function of the size or heat exchange area of the respective system (regenerator or recuperator). The thermal efficiency of any air preheater is also limited to about 75 to 80% by the ratio of the heat capacity flow (product of the specific heat capacity and volume flow rate) of air and flue gas. The flow rate and heat capacity of the air are about 10% smaller than those of the flow gas. The temperature of the flue gas leaving the air preheater is due to these general conditions in a range from 400 up 550 °C at oxygen content in the flue gas of not more than 4% after reversal system of a regenerative furnace. The dilution of the flue gas by air arising from any leakage in the system may lead to lower temperatures, but this not an improvement of the whole situation because the energy content of the flues remains unchanged in this case.

Due to this reason many glass companies were looking for a technical solution to bring the energy back into the melting process with a minimum of additional installations. There existed already different waste heat recovery concepts in the glass industry mainly based on boiler systems combined with production of electrical energy. The efficiency of boiler systems is limited due to the low temperature level of the flue gases behind regenerative glass melting furnaces which was mentioned already before. The investments for a boiler system combined with a cogeneration system are relatively high; the use of the heat for any application like heating buildings which is not directly linked to the melting process is very limited.

PROCESS IMPROVEMENTS BY BATCH PREHEATING

The idea to preheat the raw materials with the free available waste heat from the melting process was born already before, probably in the nineteen fifties or even earlier. But many problems were expected for the application of any batch preheating process, starting from chemical reactions between the flue gas, the creation of agglomerations based on reactions between the flue gas humidity and soda ash which would block the batch transport by gravity up to the increased carry over of dry batch ingredients from the combustion space into the regenerator.

Calculations carried out by the author showed an energy saving potential of 12 – 18 % by preheating the batch up to 350 °C (1). Some basic influence factors are:

- The cullet content of the complete batch mixture: because in any batch preheater only the heat content (=temperature) of the mixture should be increase but no chemical reaction should occur a high cullet proportion increases the potential energy savings of the respective.
- The humidity in the batch charged to the furnace will be evaporated in the batch preheater. At a batch preheating temperature of 350 °C, and a batch humidity content of 2% the energy for evaporating the water amounts to about 10% of energy transferred to the raw material mixture in the batch preheater
- A furnace with a high pull rate has better conditions for the installation of a batch preheater because the wall losses of the furnace itself will not by affected by preheating the batch. In terms of absolute energy savings the influence of the batch preheating will be equal, but in terms of relative energy savings of the whole process a furnace with a high pull rate has better conditions.
- In the case of a high amount of electrical boosting it is possible to save primarily parts of the more expensive electric energy.
- The preferred fuel for the installation should contain a minimum of sulphur and any other ingredient which can result in flue gas components which can react with the batch ingredients; from this point of view natural gas was the first choice.

Taking into account these difference factors container glass furnaces with a high pull rate and a high cullet amount are typical candidates for the installation of a batch preheater. Especially for the production of green glass these furnaces are operated with high cullet content and often also with a high proportion of electrical energy. The energy savings which are obtained under practical conditions depend strongly from the layout of the respective furnace and the conditions under which the furnace is operated. They should be analysed and calculated for each individual furnace separately. A detailed analysis shows that the best results will be obtained when also the possible increase of the melting rate of the furnace is utilised. Under this perspective the batch preheater should be planned as an integrated part of the complete furnace system and not as separate additional part of the installation.

ENVIRONMENTAL CONSIDERATIONS

Basically four emissions parameters have to be taken into account regarding the flue gases of glass melting furnaces; values in brackets are limits due to the legislation in 1986 (3):

- Dust (50 mg/m³)[1]
- NO_x (1,8 – 3,5 g/m³)
- SO_x (1,8 g/m³)
- Halides (HCl: 30 mg/m³; HF: 5 mg/m³)

For a batch preheating system working with direct between especially the limit on the dust concentration was expected to be a problem. Typical dust concentrations of standard glass melting

---

[1] All concentrations of dust, NO_x, SO_x and halides are given in standardized figures: all concentrations are normalized to 8% oxygen in the dry flue gas (standard for glass melting furnaces in Germany)

furnaces were in the range of $150 - 200$ mg/m^3. Due to the development in the 80s especially in Germany the installation of filter system to reduce the dust concentration in the flue gas was expected for any glass melting furnace. Generally the dust filter was combined with a dry sorption system, where calcium hydrate or soda ash could be added to take out SO_x and halides out of the flue gas system. As impact of the batch preheater on these emission parameters was an increase of the dust concentration combined with a reduction of SO_x and halides expected. A direct influence of the batch preheating process on the NO_x concentration was not expected, only a reduction of the NO_x mass flow proportional to the savings in natural gas.

CONCEPT AND DEVELOPMENT OF THE BATCH PREHEATER
The basic demands to any batch preheater can be formulated as followed:
- Minimum changes in the operation conditions in the plant
- Parallel operation of furnaces with and without batch preheater with one batch house
- No or at least a minimum of moving parts
- No additional energy consumption
- Long lifetime at minimum maintenance
- Minimum changes in the furnace layout
- Installation in existing plants should be possible

This are only some major demands, this list could be continued endless. The "Nienburger" preheater is a solution which satisfies the demands described here nearly completely. It was invented and developed by Helmut Roloff (2).

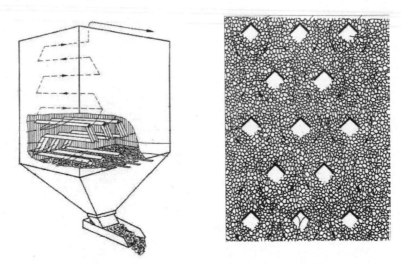

Picture 1: Basic concept of the batch preheating system "Nienburger" type
(perspective view and cross section)

The basic concept and function is shown in picture 1. The batch preheater replaces the batch hopper which is normally located above the batch charger und is supplied witch batch at the top end. The flow

of the flue gases is controlled by so called "roof" elements which are surrounded by the batch mixture. Totally there are 8 – 10 roof layers which are link between the different layers at the outer walls of the preheater. Because of the vertical flow of the batch mixture a hollow space under the roof is created. This hollow space is used to guide the flue gas through the preheater in a cross counter flow relative to the raw material. The flue gas enters the preheater at the lower end with a temperature of about 400 – 450 °C at typical regenerative furnaces and is leaving the preheater at the top, with a temperature of 250 – 300 °C.

Beginning in the nearly 80s different versions of the preheater were built which were tested and operated offline separated from the glass melting furnaces. Many improvements were introduced already in the offline trial versions, especially the connection system between the different layers which is open at the lower end and guarantees that the any dust disposed at the walls will be taken up by the material flow again. Only very limited experiments concerning the melting behaviour of the batch in a real furnace were possible. Some manual trials showed an accelerated melting of the preheated batch material. Even the basic possibility to build a batch preheating system could be shown in the offline experiment, the practical installation of a batch preheater on a running furnace was a high risk.

Even though a first practical installation was integrated at furnace #4 in Nienburg and started in December 1987.

INSTALLATIONS AT CROSS FIRED REGENERATIVE FURNACES

Considering the basic considerations with respect to the batch preheating process described before the furnace selected for the first installation of a batch preheater was an ideal choice: it was a green glass furnace operated at high pull rates with high cullet content in the mixture. Some basic data:

- Furnace type: Cross fired with regenerative air preheating
- Burner pairs: 4
- Batch chargers: 5
- Melting area: 82 m²
- Pull rate: 260 – 310 t/d
- Type of glass: Soda lime, green, >80% cullet content in the batch
- Fuel: Natural gas type L (~12 – 13 % N_2, 32000 kJ/m³)
- Boosting: 800   1400 kW

Because of direct contact between flue gas and raw materials in the preheater and due to the discussed limits in the dust concentration the batch preheater was installed together with an electrostatic precipitator (System McGill, built by GEA / Interproject).

The batch preheater itself is installed instead of the batch hopper normally located above the doghouse. In case of cross fired furnaces the batch preheater is about as wide as the furnace itself, about 2 – 2,5 m deep and 20m high (measured from the glass level). The raw material content is about 70 – 80 tons of normal batch and cullet mixture, which is supplied from the batch house with any changes or additional installations. Beside the batch preheater also a raw material bypass has to be installed. The weight of the steelwork necessary for the preheater and the preheater itself is about 120 t.

A first report about the experiences with this batch preheating system was presented in session of one of the technical committees of the German glass society (DGG) (4). The specific energy consumption of this furnace was about 3550 kJ/kg glass. A careful analysis of the energy balance of this furnace taking into account the important influence factors as electric boosting, regenerator size, drying and preheating the batch showed a good consistency of the practically obtained specific energy consumption with the calculated figures. The obtained energy saving of this furnace was about 16% after correcting for the different influence factors.

This furnace showed the practical evidence, that the successful operation of glass melting furnace with a batch preheater was technically possible and that the expected calculated energy savings can be achieved in a practical installation. The furnace was operated over 12 years with a total melting output of about 1.1 million tons over the whole campaign.

Even the first installation of this batch preheater was successful, the experience showed also the demand for some improvements. The major points were the demand for an improved batch transport to avoid the creation of any larger batch agglomerations and dust concentration in the flue gas leaving the preheater. The preheater had to be cleaned one or two times per year.

1999 the furnace was repaired and improved in many points. The melting area was enlarged from 82 to 112 m², the regenerators and all other surrounding parts of the furnace were adjusted to the new furnace layout. The static construction and the active part of the preheater remained nearly unchanged. The batch exit of the preheater was adjusted to the new charging situation (wider furnace). The flue gas channels in the upper three layers were linked to a parallel flow situation to reduce the flue gas speed drastically.

Since the restart of the new furnace 4 in Nienburg the preheater is now running without any interruption for cleaning, especially removing agglomerations. Compared to the original layout the present version is a big step forward in reliability and constant operation. The average energy consumption over the first 7 years was 3367 kJ/kg glass including the electric boosting. After correcting the electric boosting the specific energy based on the part molten by natural gas is 3648 kJ/kg. The furnace is now running since nearly nine years. Picture 2 shows the specific energy consumption of this furnace after the rebuild as a function of the pull rate for the first nine years.

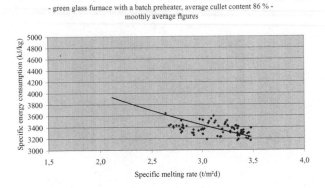

Picture 2: Specific energy consumption of green glass furnace with batch preheating as function of the specific pull rate

Another example, which will be discussed in more detail, is furnace #1 also in Nienburg. This furnace was started as a completely new furnace in 1991. Also this furnace was started with a batch preheater from the beginning. The basic data of this furnace are given in the following table:

- Furnace type: Cross fired with regenerative air preheating
- Burner pairs: 4
- Batch chargers: 6
- Melting area: 108 m²
- Pull rate: 260 – 340 t/d
- Type of glass: Soda lime, flint, 40 - 70% cullet content in the batch

Due to the production of flint glass the content in the mixture of this furnace was lower than under green glass conditions. In the first 2-3 years the cullet content was generally below 50%, due to the improving supply with flint cullet from the market in later years the cullet content could be raised to about 60 %. This furnace was operated until 1995 with a small interruption for a hot repair in 2004. In 2005 the furnace was rebuilt with changing the basic layout. Only minor changes were introduced in the furnace construction. The batch preheater was optimized based on the experience available from the green glass furnace mentioned before.

Picture 3 shows the specific energy consumption of this furnace as a function of the specific pull rate based on the data from the first 3 years after the repair. In principle the results are similar to those of furnace #4 Nienburg. The average specific energy consumption is 3870 kJ/kg, 6 % higher compared to the green glass furnace described before. But this furnace is operated without boosting, at lower pull rates and with lower cullet content than the green glass furnace. Just the lower cullet (- 24%) is sufficient to explain the higher specific energy.

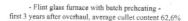

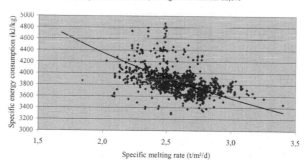

Picture 3: Specific energy consumption of flint glass furnace with batch preheating as function of the specific pull rate

A second flint glass furnace with a batch preheater was built in 1995 in Neuenhagen. This technical data are of this furnace are in principle the same like those of furnace #1 Nienburg mentioned before.

The first batch preheater installations have been made on cross fired regenerative furnaces. There are two reasons which make the installation on cross fired furnaces easier:

- Cross fired furnaces can be charged using one big doghouse over the complete width of the furnace. This guaranteed a constant batch flow over the whole width of the batch preheater.
- At cross fired furnaces there is always an area with lower air and flue gas speed between the first burner and the doghouse. Due to the heat radiation from the inner parts of the furnace into the direction of the doghouse the batch has already a molten surface with a thin layer of molten glass on the upper side, when it reaching the region with appreciable gas speeds. This behavior results in a lower dust concentration in the flues entering the regenerator.

END FIRED AND OXY FUEL FURNACES

The charging situation at end fired furnaces is totally different to the situation at cross fired furnaces with one big dog house.

- The complete raw material flow is charged into one single doghouse with a small surface area and a small opening into the furnace. In some cases there two doghouses, but even then the basic situation is unchanged.
- After entering the furnace the mixture is immediately in direct contact with the flow of preheated air or flue gas leaving the combustion space in direction to the regenerator. The carry over of raw materials inside the furnace and to the regenerator chambers is favored compared to cross fired furnaces.

Due to these differences the installation of a batch preheater on an end fired furnace is on the first view more complicated and combined with a higher risk. In 1992 the first batch preheater of the layout shown was installed at furnace #6 in the plant Wahlstedt.

- Furnace type: End fired with regenerative air preheating
- Doghouse: One at the left side
- Melting area: 70 m²
- Pull rate: 200 - 240 t/d
- Type of glass: Soda lime, flint and green , 40 - 85% cullet content in the batch
- Fuel: Natural gas type H (36000 kJ/m³)
- Boosting: Mainly used for green glass

Even the energy consumption and the melting behavior was similar, the first solution selected for the batch transport from the furnace to the batch preheater was not satisfying. The preheater consisted of two parts with a quadratic ground plot, which were arranged side by side over the doghouse. Under each part of the two parts a big round vibrating chute was installed. From the two chutes the batch was transported to the charging machine with two slides inclined by about 45°. Many blockages characterized the batch transport during the first year.

The batch transport problem at this furnace was solved by the installation of screw conveyors under the batch preheater. Under each quadratic part there were 5, in total 10screw conveyors. The link to the charging machine consisted of two additional screw conveyors collecting the material flow coming from the 10 first mentioned screw conveyors. With this installation the transport problem was solved with sustainable success. The furnace was operated for about 13 years for the production of green and flint glass. In 2005 the furnace was closed due to market reasons.

Another batch preheater of the "Nienburger" type was installed in 1997 behind an oxy fuel furnace at the former BSN plant in Düsseldorf (5). This furnace was used for flint glass only and shut down due to market reasons in 2005 together with the complete site Düsseldorf.

The following table gives an overview of all "Nienburger" batch preheater installations.

Plant, furnace	Type of furnace	Year of installation	Glass	Remark
Nienburg #4	Cross fired, 4 burner pairs	1987	Green	Overhaul & improvement 1999
Nienburg #1	Cross fired, 4 burner pairs	1991	Flint	Overhaul & improvement 2005
Wahlstedt #6	End fired	1992	Flint & green	Furnace shut down due market reasons
Neuenhagen #9	Cross fired, 4 burner pairs	1995	Flint	Overhaul & improvement 2006
Düsseldorf #2	Oxy fuel	1997	Flint	Furnace shut down due market reasons

ENVIRONMENTAL BEHAVIOR

Generally the batch preheating process makes energy helps to reduce any type of energy related emission at least proportional to the obtained energy saving. The first and in the last years most important parameter influenced by these energy savings is the NO_x emission of the respective furnace. Together with different improvements of the burner layout the NO_x concentration of the cross fired furnace mentioned here could be reduced from $6 - 7$ kg NO_x/ton to less than 1 kg NO_x/ton. In terms of the NO_x concentration in the flue gas this was a reduction from $3000 - 4000$ mg/m³ to less than 800 mg/m³.

Due to general development in Germany and neighbouring countries the concentration of dust in the flue gas of glass melting furnaces was reduced from originally 150 mg/m³ to 50 mg/m³ in 1986, meanwhile to 30 or 20 mg/m³ depending from the respective individual installation. Also furnaces without a batch preheating system have a dust concentration of $150 - 200$ mg/m³ and even more. Because also the capture of SO_x as well as halides is necessary to meet the limits according to the legislation, a sorption reactor where $Ca(OH)_2$ or Soda is added must be installed. The dust must also be removed using appropriate filter systems. Due to all these influence factors electrostatic precipitators or fabric filters became standard equipment in 80s and 90s.

Because a batch preheater with direct contact between flue gas and raw material must lead to an increased dust emission, the obligatory presence of dust filter systems behind glass melting furnaces was an additional precondition for the installation of batch preheating systems. All batch preheaters mentioned in this paper were combined with filter systems from the beginning.

Typical figures for the concentration of dust in the flue gas of a glass melting furnace with a batch preheater measured at different sampling points in the flue gas system are given in the following table.

Concentration of dust along the way of the flue gas				
Sampling point	O2 [%]	Flue gas [m³/h]	Dust [mg/m³]	Dust [mg/m³, 8% O2]
Before batch preheater	3,2	14900	96,5	132,1
After batch preheater	11,2	27070	1675,0	2270,0
After precipitator	12,7	31963	22,4	35,0

The dust concentration in the flue gas after leaving the precipitator is similar to furnaces without a batch preheater, where $Ca(OH)_2$ or Soda must be added to the flue gas before the precipitator.

The next table shows as an example the concentration of SO_x, HCl and HF in the flue gas of a glass melting furnace with a batch preheater.

Concentration of HF, HCl and SO2 at different sampling points			
Sampling point	SO_x (as SO_2) [mg/m³]	HCl [mg/m³]	HF [mg/m³]
Before batch preheater	~2000	91,0	18,0
After batch preheater	816	9,5	5,0
After precipitator	641	2,4	0,5

The results shows, that all limits of the present legislation can be kept with the existing batch preheating systems combined with an electrostatic precipitator.

Mean while the use of sodium sulphate as refining agent was reduced in different furnaces with a batch preheater, because a big part the SO_x, HCl and HF contents in the flue gas is already reduced in the batch preheater. Under optimum conditions furnaces with batch preheating can be operated without adding $Ca(OH)_2$ or Soda.

CONCLUSION

Twenty years after the first batch preheater according to the Nienburger has been installed layout at a glass melting furnace this technology shows a high degree of reliability. The practical energy savings compared to glass melting furnaces without a batch preheater are in the range of about 15 % and agree with the expectations based on mathematical models. The batch preheater must be integrated as part of the complete furnace concept also including the environmental systems like electrostatic precipitator.

Based on the energy savings the batch preheating is also a contribution to improve the environmental situation of the glass melting process.

Until now all known installations of the preheater system are operated in combination with container glass furnaces.

LITERATURE REFERENCES

1. Barklage-Hilgefort H., Trier W.: Berechnungen zum Einfluss der Rohstoffvorwärmung auf den Wärmehaushalt von Glasschmelzöfen. Glastechn. Ber.**56** (1983) p. 269 - 279

2. Roloff H.: Process and apparatus for preheating raw material for glass production, especially cullet. European Patent EP0211977. Publication date August 14, 1985

3. TA Luft „Vorschriften zur Reinhaltung der Luft", Version dated on February 27, 1986

4. Enninga, G.: Betriebserfahrungen mit dem Rohstoffvorwärmer der Firma Nienburger Glas. Presentation at the meeting of the Techn. Committee II of the DGG (Furnace construction). Oktober 18, 1988, Würzburg/Germany

5. Hufen, D. Erste Betriebserfahrungen mit einer sauerstoff-brennstoffbeheizten Behälterglaswanne mit vorgeschaltetem Scherben-/Gemengevorwärmer. Presentation at the meeting of the Techn. Committee VI of the DGG (Environmental items). Oktober 15, 1997, Würzburg/Germany

ENERGY SAVING OPTIONS FOR GLASS FURNACES & RECOVERY OF HEAT FROM THEIR FLUE GASES AND EXPERIENCES WITH BATCH & CULLET PRE-HEATERS APPLIED IN THE GLASS INDUSTRY

Ruud Beerkens
TNO Science & Industry
Eindhoven, The Netherlands

ABSTRACT
Several measures, such as changes in batch composition (less batch humidity), or optimization of operating conditions, and limiting the combustion air excess, can lead to typically 2-8 % of energy savings of industrial glass furnaces. Larger energy savings are only possible by new furnace designs, more insulation, increased cullet ratios or extra recovery of the heat contents of the flue gases. The flue gas heat contents, downstream regenerators or recuperators of air-fired furnaces or downstream the exhaust of oxygen-fired glass melting furnaces, can be exploited to preheat batch and cullet up to 275-350 °C for regenerative furnaces and up to more than 500 °C for recuperative and oxygen-fired furnaces. Above 550-600 °C, the batches may start to show sticking behaviour.
Energy savings of 12-20 % have been reported for regenerative air-fired glass furnaces after connecting a batch and pellet-preheat system. The highest savings in the consumption of specific energy (energy consumption per metric ton molten glass), can be achieved by combining the application of batch and cullet preheating with an increased pull rate. Increased pull rates of more than 10 % have been achieved and this potential of increasing production in the same furnace will improve economics of batch / cullet pre-heaters. Almost all batch or cullet pre-heater applications are found in the container glass industry. The pay-back of the capital costs of batch & cullet preheating systems by energy cost savings may take more than 3-4 years, depending on the energy prices and modifications required to the batch and flue gas channel systems.
There are 5 or 6 different types of batch and/or cullet preheating systems applied in the glass industry or still in a testing phase. All pre-heaters deliver very dry batch or cullet. This dry batch charged into the furnace may cause batch carry-over (depending on the doghouse design and position of the burners relative to the batch blanket moving from the doghouse into the tank). Especially in end-port fired furnaces, this dust formation in the furnace may lead to fouling of the regenerator checkers.
Other problems that have been reported are: odour in case of high levels of externally recycled cullet (with organic contamination) and dust formation and dust deposition in the ambient space surrounding the furnace, especially in the case of long distances between pre-heater and furnace.
Direct contact pre-heaters (direct contact between the flue gases and the batch, to be preheated) will show acid gas scrubbing potential. Batch components absorb SOx, HCl, HF and selenium compounds from the flue gas stream.

ENERGY BALANCES OF FURNACES EQUIPPED WITH AND WITHOUT BATCH PREHEATING SYSTEMS

The energy contents of the flue gases (downstream of the regenerators) of modern end-port fired regenerative furnaces, typically constitutes 30-35 % of the supplied energy to the furnace. About 40-48 % of the energy input in these furnaces is used for heating the batch and melt, for endothermic fusion reactions and water evaporation. The residual amount of energy is lost by heat conduction through the furnace walls, forced cooling, leakage of hot gases or radiation heat losses.

In the case of all oxygen-fired glass furnaces (smaller flue gas volume flows, but higher temperatures compared to regenerative air-fired glass furnaces), about 30-35 % of the combustion heat is lost by the flue gases. Part of these flue gas heat losses can be recuperated by preheating cullet or batch plus cullet. The preheat temperature is determined by the flue gas temperature, heat contents of the flue gases and

the preheating system dimensions and design. Generally in a counter flow heat exchanging system, the heat transfer from flue gas to batch is more efficient than in a co-flow system.

The flue gas volume flows of all oxygen-fired (oxygen-natural gas or oxygen-fuel-oil) are much lower (typical 3.5 to more than 5 times lower) compared to the volume flows of the flue gases from regenerative furnaces with similar glass melt pull. Temperatures before dilution or quenching are much higher for the oxygen fired furnaces. These higher temperatures enable a relatively high batch and cullet-preheat temperature [1].

Typical batch/cullet or cullet preheat temperatures for regenerative furnace conditions are 275-325 °C, but for oxygen-fired furnaces preheating up to 450-500 °C seems feasible with the available temperature levels and flue gas heat contents. Most batches do not react or start to show sticking behaviour, at and below these temperature levels. TNO tested sticking behaviour of float glass raw material batches and these tests showed no sticking below 550-600 °C.

Concerning the energy balance of a glass furnace; the preheating of batch means that extra heat (heat contents of batch) is supplied to the glass furnace by this batch and consequently less fuel is required for melting. Less fuel consumption will result in lower volume flows of the combustion gases and flue gases, therewith reducing the flue gas heat contents and heat losses. This lower volume flow of flue gases and the recuperation of part of these flue gas heat contents (by preheated batch) both will lead to energy savings. In addition to this, the preheating of batch or cullet at temperatures far above 100 °C will eliminate the water from the batch (typically batch and cullet may contain 2-4 mass-% of water). Water evaporation takes place in the pre-heater system and not anymore in the glass furnace. Thus, the extra fuel required in the furnace to evaporate water from the batch in the furnace can be saved and also the water contents of the flue gases exhausted by the furnace will decrease, leading to fuel savings.

A complete energy balance calculation, based on heat required for melting and reactions of the batch, heating of the glass, structural heat losses, heat of combustion, energy supplied by preheated batch and electric boosting and heat losses by evaporation processes or flue gases (depending on flue gas volume flow, composition of flue gas and temperature) will give the effect on the energy consumption changes when batch and cullet is preheated plus dried by the heat contents of the flue gases.

Figure 1a shows a typical energy balance of a modern end-port fired container glass furnace, equipped with regenerators (existing installation) without pre-heater.

Figure 1b shows the energy balance of the same furnace, but now equipped with a batch & cullet preheating system with batch temperature of 300 °C. Energy savings with unchanged pull are about 15 %. The achievable energy savings, depend very much on the water content in the batch and the energy efficiency of the furnace without batch, in case of high structural energy losses, the energy savings (expressed in %) after batch preheating will be less compared to a well insulated furnace.

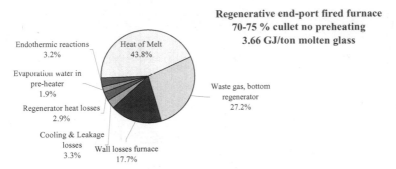

Regenerative end-port fired furnace
70-75 % cullet no preheating
3.66 GJ/ton molten glass

Figure 1a   Distribution of supplied energy in energy-efficient (top ranking in energy   efficiency benchmark) regenerative container glass furnace

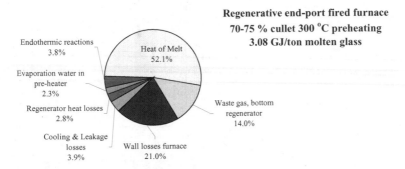

Regenerative end-port fired furnace
70-75 % cullet 300 °C preheating
3.08 GJ/ton molten glass

Figure 1b Distribution of external supplied energy in energy-efficient (top ranking in energy efficiency benchmark) regenerative container glass furnace equipped with batch & cullet preheating system (% given as fraction from energy input by fuel excluding energy input by the preheated batch)

Table 1 shows per ton molten glass the energy distribution among endothermic reactions, heating of melt, wall losses, cooling losses, flue gas heat (through chimney), water evaporation for the same furnace, with and without batch / cullet preheating (for batch preheating case, estimated by energy balance modelling).

Not only the extra heat input by the preheated batch will lower the energy demand for the glass furnace, but also the flue gas volume flows will decrease (thus decreasing the flue gas heat losses) and no heat for evaporating the water from the batch in the glass furnace is required. Typical energy savings depend on water content of the non-preheated batch and cullet% and batch preheat temperature. Energy savings for very well insulated furnaces are typically 12-18 % or even higher per ton molten glass in case of batch/cullet preheating in combination with a pull increase: 18-22 % specific energy

savings. In case of badly insulated furnaces, these energy savings expressed in % of the original energy consumption will be a few percent less.

Table 1   Energy balance of regenerative furnace with and without batch pre-heater
          This furnace is among the most energy efficient according to the energy
          efficiency benchmark studies [2].

Energy balance of modern end-port fired regenerative container glass furnace			
Pull	255	metric tons glass per day	
Cullet%	70	based on glass mass	
Batch humidity	3	mass% of batch	
Batch preheat temperature	300	°C	
ENERGY BALANCE			
OUTPUT	unit	without preheater	with preheater
Evaporation of water	MJ/ton* molten glass	72	0
Endothermic reactions melting process	MJ/ton molten glass	131	131
Heat enthalpy of melt (throat)	MJ/ton molten glass	1637	1637
Flue gas heat content (exit regenerator)	MJ/ton molten glass	1052	783
Conduction of heat through furnace walls	MJ/ton molten glass	661	661
Conduction of heat through regenerator	MJ/ton molten glass	119	119
Cooling & leakage heat losses	MJ/ton molten glass	87	87
INPUT		without preheater	with preheater
Fuel combustion	MJ/ton molten glass	**3675**	**3023**
Air (before preheating)	MJ/ton molten glass	35	28
Batch	MJ/ton molten glass	24	317
Latent heat natural gas	MJ/ton molten glass	4.2	3.4
Flue gas volume flow into regenerator	Nm³/hr	13760	10647
Heat enthalpy flue gas into regenerator	MJ/hr	30760	24182
Energy savings	MJ/ton molten glass		**652**
Nm³ = 1 m³ at 273.15 K, 1013 mbar			
* 1 ton = 1000 kg			

As previously indicated, the exhaust gases from all oxygen-fired furnaces not passing a regenerator first, will have a much higher temperature and enable higher batch/cullet preheat temperatures compared to regenerative air-fired furnaces.

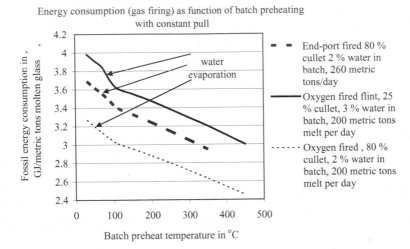

Figure 2   Effect of batch & cullet preheat temperature on energy consumption of a glass furnace according to energy balance modelling. Energy consumption includes only combustion energy, excluding energy for oxygen production.

Energy balance modelling shows for 450 °C batch / cullet preheating a potential energy and oxygen savings of 20 to more than 25 % of full oxygen-gas fired container glass furnaces melting about 200-250 tons molten glass per day, the higher values (25 % and even higher) referring to about 10 % increased pull rates in combination with batch/cullet preheating. In general, it can be concluded that batch plus cullet preheating up to 500 °C and the same pull, will gain 1 GJ energy savings per ton molten container glass (almost independent of the baseline energy consumption without batch preheating). Figure 2 shows the effect of batch pre-heating (and drying) for two different container glass furnace types on the energy consumption (excluding the energy demand for oxygen generation). For an oxygen-fired glass furnace, two situations are depicted: one case operating with 25 mass-% cullet (3 % water in the batch) and one situation with 80 mass-% cullet in the batch, with only 2 % water. Comparing the regenerative end-port fired furnace (260 metric tons/day) with the oxygen-fired furnace, for the same cullet level and batch humidity, first of all, one observes from the figure that the oxygen-fired furnace (200 metric tons/day) consumes less energy even at a lower pull. However, the energy consumption for oxygen generation is not taken into account in figure 2.

Secondly a higher energy saving potential by batch preheating is expected for the oxygen-fired furnace type, because of the higher preheat temperatures that are possible for oxygen-fired furnaces, with their higher flue gas temperatures.

ENERGY SAVING OPTIONS FOR GLASS FURNACES
Energy balance models are able to predict the effect of changes in process parameters, furnace design and batch composition on the energy demand of a glass melting process in industrial glass furnaces.

The effect of individual changes in furnace operation on energy consumption are difficult or almost impossible to access from industrial measurements, because <u>average</u> energy consumption has to be measured for a longer period of time (several days) and in that case more than only one of the process parameters have generally been changed.

Furnace design parameters: The furnace size and glass melt surface area with respect to the melting load (often expressed in metric tons per day, see figure 3) and specific melting rate (metric tons/m² per day) will have an effect on the energy consumption. The type of furnace: electric melter, recuperative furnace, regenerative fired furnace (end-port or cross-fired) or directly oxygen-fired melter will be important for the specific energy consumption as well. Benchmark studies [2] show that in the container glass industry, the end-port fired regenerative furnaces and the latest generation oxygen fired furnaces are the most energy efficient furnace types when considering the total energy input including the primary energy required to generate electricity and oxygen. End-port fired furnaces with regenerators showing about 70 % of heat transfer from the flue gases to the combustion air are among the most energy efficient with energy consumption levels lower than 3.5 GJ/metric ton molten glass in case of high cullet-% in the batch. Insulation is another important design aspect, today the total energy required to compensate for structural heat losses can be lower than 0.8 GJ/metric ton molten glass for highly loaded furnaces. In float glass furnaces with a longer residence time of the glass in the tank, structural heat losses above 2 GJ/metric ton molten glass can be found.

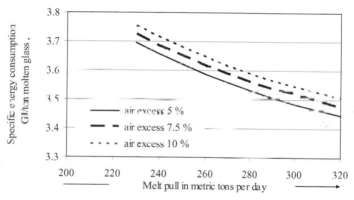

Figure 3 - Energy consumption versus pull and air excess (preheated air) for an end-port container glass furnace. At very high pull rates, specific energy consumption may increase again or glass quality problems may occur.

Batch parameters: The most important batch parameters, that determine the energy requirement for melting in glass furnaces, are the cullet fraction in the batch, water content and the contents of carbonates. In modern end-port fired container glass furnaces an increase of 10 % of the amount of cullet instead of normal batch leads to 1.5-3 % energy savings, dependent on the furnace type.
In end-port fired regenerative furnaces for container glass production, 1 % increased water content in the batch may lead to 1.5-2 % increased energy consumption (assuming no change in flue gas temperature), due to the extra heat required for evaporating water from the batch and the extra heat content of the water in the flue gases. The effect of a decrease of the water content in the batch or increasing the cullet ratio (less endothermic reactions and less CO_2 evolution) in the batch on energy

savings are even larger for oxygen-fired furnaces, because a lower amount of CO_2 and water vapour in the flue gases of oxygen-fired furnaces, with typically temperatures of 1350-1450 °C, will decrease flue gas heat losses more than for the lower CO_2 and water vapour flows from regenerative furnaces with flue gas temperatures downstream the regenerators of about 500 °C. See table 2.

Table 2: Effect of batch humidity (based on total batch) & cullet fraction in batch on changes in energy demand in glass melting processes according modelling. In the calculations it is assumed that the flue gas temperature and melting kinetics is not changed when increasing the water content in the batch.

	End-port fired regenerative, 80 % cullet, 2 % water in batch, 260 metric tons/day	Oxygen fired , 80 % cullet, 2 % water in batch, 200 metric tons melt per day	
1 % more batch humidity	1.83	2.4	% extra energy consumption
1 % more batch humidity	0.067	0.08	extra energy GJ/ton melt
10 % less cullet	2.4	3.2	% extra energy consumption
10 % less cullet	0.0875	0.11	extra energy GJ/ton melt

Replacing carbonates (e.g. $CaCO_3$) for the oxides (CaO) will reduce the energy demand drastically because the decomposition of carbonates in the batch blanket is a endothermic reaction with high energy demands. Example: replacing all limestone by quick lime in a soda-lime-silica container glass forming raw material batch with 25 % cullet delivers an energy savings of about 4-5 % (no dolomite in batch).

Process parameters: Other important aspects for the energy consumption (and CO_2 emissions) of industrial glass furnaces are the oxygen excess in the combustion process. See figure 3. For a float glass furnace a decrease of the O_2 content in the exhaust gas (top regenerators) from an average value of 2 vol.-% down to 1.4 vol. % will give 1.6 % energy savings. Especially the infiltration of non-preheated air will cause very high extra energy demands. For a 300 metric tons glass melting furnace, infiltration of 600 Nm^3 cold air (about 5 % of the total air demand) causes 5 % extra energy consumption. Control of the air excess is very important to limit NOx formation in the combustion chamber, but also to avoid too high fuel demands for melting. The creation of soot in natural gas flames is another important parameter that will decrease the energy consumption in a glass furnace, because of an improved net heat transfer from the flames into the melt. It is expected that an increase from an overall grey emissivity from 0.17 to 0.25 in a flame decreases the energy consumption for a regenerative furnace with at least about 2.5-3 %.
Control of the glass melt temperature in the throat may lead to moderate energy savings, typically 0.5 % if the average glass melt temperature can be decreased by 10 °C (18 F).
The flue gases of the glass furnaces, even downstream regenerator typically contain 30-40 % of the energy input of the glass furnace. Flue gas heat can be used for steam production by waste heat boilers. This steam can be used to generate electricity. In Germany a few glass furnaces have been equipped with a steam-turbine-regenerator system. Pay-back times however are considered to be too high in most cases.
In the next sections the most effective way of decreasing the energy consumption of glass furnaces, the partly reclaim of the heat contents of the flue gases by batch & cullet preheating will be discussed.

DIFFERENT CULLET AND BATCH/CULLET PREHEATING SYSTEMS APPLIED IN THE GLASS INDUSTRY

At present, there are 5 different cullet or batch/cullet preheating systems that have been applied or developed for the glass industry. To the authors' knowledge, 7 installations are currently in operation. Six more have been operating in the past. In total 13 full scale systems have been build, 12 in Europe (mainly Germany) and 1 in he USA.

The different systems can be distinguished in:
   a. Systems for cullet preheating only;
   b. Systems for batch plus cullet. There have been no experiences with only normal batch (without cullet) preheating. Most batch/cullet preheating systems work with > 50 mass% cullet;
   c. Systems with physically separated channels or shafts for batch plus cullet on one hand side and flue gases on the other;
   d. Systems with direct contact between the flue gases and material to be preheated.

The different existing systems and some new developments are shortly summarized in this section.

The SORG cullet pre-heater system [3-5]:

Figure 3 shows a sketch of this system applied at two recuperatively fired container glass furnaces (SORG LONOX® melters) in Germany.

* The system is applied since 1987 in Germany. The first installations are still in operation.
* It preheats only cullet (internal and external: post consumer), and it is preferably used for glass furnace with high cullet-fractions in the batch, example 85-90 % cullet.
* The use of clean cullet or aged cullet from which most organic materials (food residues) are already removed by natural fermentation processes is preferred to avoid odour/fume problems.
* Investment for 350-400 metric tons cullet per day: ±2.5 million dollars (2007).
* Cullet preheating: typically ±300 °C, depending on flue gas temperatures and heat contents. Typical flue gas inlet temperature: 550-600 °C.
* The fraction of fine cullet < 3.5 mm should be limited (recommendation < 10 %)
* Residence time of cullet in preheating system: 4 hours.
* Size of pre-heater: 30-50 tons cullet hold-up.
* Heat exchange area per pre-heater 65-75 m^2 (for 200-300 tons cullet/day)
* Typical dimensions per module: Height: 6 meter,   Width: 3 m   Depth: 0.4 m
* Applied to oil and gas fired recuperative glass melting furnaces (SORG LoNOx® types): specific energy consumption < 850-900 kcal/metric ton glass (3.35-3.58 MMBTU/short ton)

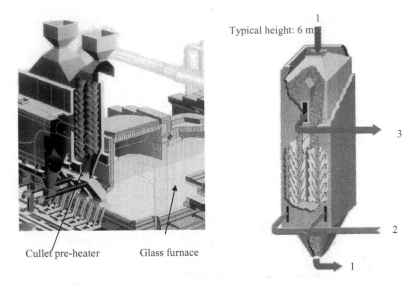

Figure 4   The SORG cullet preheating system Cullet (1) moves from top through modules to bottom and flue gases coming from the recuperator(s) flow through the cullet bed and special   steel wings from bottom (red arrows (2)) to the top (blue arrow(3)). Flue gases cool down, typically from 550 to 250-300 °C.

- Estimated energy savings by cullet preheating: 15-20 % (recuperative furnace): from 4.22 GJ/ton (4 MMBTU/ton) to 3.41 GJ/ton (3.24 MMBTU/ton), based on 85 % cullet.
- 3 Systems have been installed at recuperative furnaces for furnaces > 65 % cullet, 2 are still in operation in the container glass industry (both in Germany).
- Pressure drop (flue gas) over cullet column: ±3 mbar.
- Extra fan power required (for furnace pull 350-400 tons per day) to compensate for pressure drop: 10-15 kW (estimated for 22000 Nm³ flue gas/hr).

The PRAXAIR-Edmeston Cullet Preheat system, with electrostatic precipitation in one of the cullet preheat modules [6-8]: Figure 5 schematically shows this system, applied to an oxygen-natural gas fired container glass furnace.

There is one application still in operation. One similar (previous generation version) has been applied at an air-fired glass furnace in Europe (Ireland) in the container glass industry, but the glass plant has been stopped operation.

Flue gases from an oxygen-gas fired glass furnaces are diluted with air to cool down (550-600 °C), and conducted by a fan (at lower temperature side of the system) into the cullet preheating system. The flue gases are split into two flows. One portion of flue gas is flowing through a so called pyrolyzer, where external cullet (often with organic contamination) is heated by the hot flue gases and the organic vapours or their reaction products are combined with these flue gases of the glass furnace, returning

back to the combustion chamber of the glass furnace for complete combustion of the organics. Another portion of the flue gases from the furnace, is partly quenched by cold (ambient) air and flows first into an ionizer, where the primary dust particles (mainly sodium sulfate) are electrically charged. The flue gases and charged dust particles subsequently flow through the "combined EP filter-clean cullet pre-heater system". In this module, the internal (own) cullet plus optionally* (cleaned) cullet from the pyrolyzer is heated by the flue gases and in this module an electrostatic field is imposed (by a high voltage cage) from the inner axis to the outer shell of the pre-heater. Flue gases with dust particles flow from the centre axis (inner louvered wall) to the outer shell (outer louvered wall), through the cullet bed (the cullet moving from top to bottom) and the charged dust particles are collected by the electrostatic forces on the surfaces of the cullet pieces. The preheated cullet leaves the system from the bottom part of this filter and cold normal batch is added to this cullet before charging the cullet to the furnace.

The cullet preheated in the pyrolyzer can also be charged directly into the furnace doghouse.

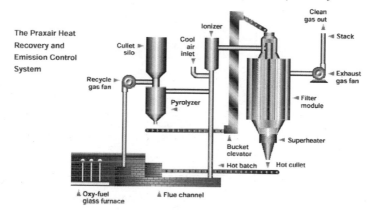

Figure 5   The PRAXAIR-Edmeston Batch preheating and integrated flue gas de-dusting system applied to an oxygen-gas fired glass furnace [6]

Some features of this PRAXAIR-Edmeston system are:
- In combination with oxygen firing, for a furnace of about 50 % cullet, energy consumption as low as 3.0-3.1 MMBTU/short ton (3.2 GJ/metric ton molten glass) has been measured [6].
- The residual particulate emissions are about 50 mg/Nm³ (8 % O_2, dry flue gas conditions) or 0.07 kg dust/metric ton glass.
- In practice the cullet is dried and preheated to 300-350 °C.
- Investment costs in 1998: 1.4 million US dollars for about 125 metric tons cullet per day.
- Operating since September 1998 in the USA in connection to a container glass furnace. *A previous version of such a system was operating in Ireland, Europe, but this glass production plant has been closed.*
- The energy consumption of the furnace hardly shows an increase in the course of time, ageing effects are very small.

- By applying the pre-heater, the pull could be increased by about 30 %, almost without increasing the fuel input.

The Interprojekt-Nienburg system (*Nienburg Glas is now affiliated to Ardagh Glass) for direct contact batch & cullet preheating by flue gases [9-13].
Figure 6 shows schematically the principle of this batch plus cullet preheating system.
Flue gases flow directly, from the regenerators or after dilution with cold air from an oxygen-fired furnace, into specially shaped flue gas channels of the preheating systems. These channels conduct the flue gases at different levels in horizontal direction through a bed of cullet mixed with normal batch.

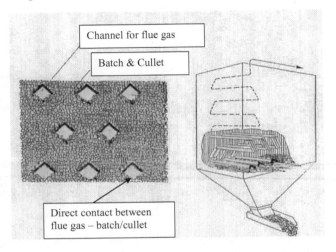

Figure 6   Principle of the batch preheating system of Interprojekt-Nienburg (Ardagh Glass) for batch plus cullet.

The flue gases are flowing at different heights (zigzag) in cross and counter flow through the downwards moving batch plus cullet. The channels are open at the bottom side (see figure 6), with a direct contact between batch/cullet and flue gases. It appears that the acid gas components in the flue gases: HCl, HF, SOx, SeO$_2$ are partly absorbed by the alkali and earth-alkali compounds in the batch (soda, limestone, dolomite). Thus the pre-heater also functions as a scrubber.
The direct contact between the flowing gases and batch causes entrainment of dust in the flue gases, up to more than 1000 mg/Nm^3 [9]. Therefore an efficient de-dusting system (in most cases an electrostatic precipitator) is always necessary to limit the particulate emissions. But, the entrained fine dust in flue gases the pre-heater will reduce the fraction of very fine dust entering the glass furnace and thus diminishes the carry-over in the glass furnace itself. In the latest generation systems (operating since 1999), the dimensions of the flue gas channels have been modified to decrease the gas velocities and momentum of the gas flow in the pre-heater in order to limit carry-over/entrainment of dust in this pre-heater device.

The most important features are summarized here (detailed information is presented in the paper of Dr. Barklage-Hilgefort [12]):

- Energy savings between 15-20 %, with batch & cullet preheating at 300-325 °C.
- 3 Systems are still in operation at regenerative container glass furnaces, all applied to cross-fired furnaces.
- 2 Systems have been installed (one for an end-port fired furnace and one for an oxygen-fired glass furnace), but are not operating anymore, because of closures of the glass production plant or furnace. One system was installed at an oxygen-gas fired glass furnace from 1997-2005, the glass plant stopped operation because of market reasons. One end-port fired furnace equipped with this type of batch & cullet pre-heater was stopped, also due to market reasons, after 13 years operation.
- Lifetime of system, at least 12 years.
- Maintenance is rather low, fouling problems are under control.
- All systems that have been built are/were in Germany.
- About 50 % of the acid gases are removed by absorption of HCl, HF and SOx in the batch. Selenium compounds in the flue gases are absorbed up to 80 % by the preheating batch and reduces the required (expensive) selenium additions to the flint glass forming batch.
- Experiences with 40-90 % of cullet in the batch.
- Energy consumption (including boosting) of a container glass furnace using about 70 % cullet in the batch and this type of batch pre-heater at 293 metric tons glass melt/day increased from 3.35 (3.2 MMBTU/short ton) to about 3.75 GJ/ton molten glass over a period of 7 years.
- Energy consumption of the large oxy-fuel fired container glass furnace (about 400 metric tons molten glass per day), using about 50-70 % cullet and equipped with a batch & cullet pre-heater was as low as about 3 GJ/ton molten glass (excluding energy required for oxygen generation) [13].
- Reported pay-back times (2007 situation): 3-4 years (depending on fossil energy price levels), excl. costs of filter system.
- Doghouse design has been adapted to avoid carry-over of the dry batch after charging this batch plus cullet in the furnace.
- Interprojekt Nienburg developed two slightly different batch & cullet pre-heating systems, one adapted for end-port fired (charging batch to the sidewall doghouse(s)) and another design for cross-fired regenerative furnaces (charging batch at the front end of the furnace).

The indirect batch plus cullet preheating system from Zippe [14, 15].
Three systems (first generation) have been built in the past in the container glass industry. One system is still in operation and this since 1996.
Figure 7 shows schematically this concept. Batch plus cullet is charged on the top of the pre-heater and moves downwards (about 1 to 3 meters per hour) through vertical shafts to the conveyor, transporting the preheated batch plus cullet mixture to the doghouse of the furnace. Most of the flue gases from the regenerators (or from recuperator) enter the preheating system in the bottom section of this equipment. Part of the hot flue gases however is used to dry the incoming (wet) batch plus cullet in the top section of the system. In this top section, the vapours (steam and organic components) have to be released without condensation in colder zones, the vapours/fumes can be added to the hot flue gases to avoid condensation. The design and dimensions of the drying section (top section) is of great importance to avoid plugging of the raw material shafts of the pre-heater. The released fumes may contain organic components, which in some cases need to be incinerated.
The flue gases entering the bottom part of this pre-heater flow through separate channels (channels separated from the batch & cullet) in the bottom modules, in horizontally direction through the pre-heater and then from one level to the next upper level (counter flow compared to batch & cullet). The flue gases flow in cross flow-counter flow direction compared to the moving batch plus cullet. The

heat transfer from the flue gases to the batch takes place through the steel separation walls. The flue gases typically cool down from 500 °C to 250 or 200 °C, pre-heating batch between 275-325 °C (depending on the heat content and temperature of flue gases). The pre-heater is preferably installed directly above the doghouse of the furnace to avoid heat losses and dust formation during transport of the dry preheated batch. It is recommended to adapt / modify the doghouse or furnace design to avoid direct contact of the batch blanket (before sintering or melting) with high-velocity hot gas flows in the furnace.

The most important features of the first generation Zippe batch & cullet preheating system are:
- No contact between flue gases and batch.
- 4 installations have been applied in the European container glass industry, 1 is still in operation: age 12.5 years, in the Netherlands. Another one was in operation from 1992-2005 in Germany at a container glass furnace. The first one was in operation in Switzerland [14] starting in 1991 at a recuperative container glass furnace (the plant closed operation). One was in operation for basalt-stonewool production.
- According to Zippe [14], batches with 50-80 % cullet can be handled. Batch plus cullet (experiences > 60 % cullet) is preheated, however with cullet lean batches (< 50-60% cullet) plugging problems have been experienced.
- Typical height of pre-heater (top): 20-25 m.
- Typical capacity: 12 tons per hour but system can be adapted to other capacities because of the modular system.
- The width of the batch channels is about 100 mm and the flue gas channels are 65-80 mm wide.
- Gas velocities 6-8 m/s, pressure drop about 500 Pascal.
- Batch plus cullet typically moves with 1 m/hour by gravity.
- Total investment costs in 1996: about 1.75-2 million US dollars (excluding incinerator for odours) [15].
- Preheat T: 260-325 °C. Flue gases typically cooling from 500 °C down to 190-200 °C.
- Energy savings (based on primary energy), due to batch & cullet preheating are about 12-15 %. In combination with an increasing pull, savings of 15-20 % have been found for pre-heating the complete batch in an end-port fired furnace in the Netherlands.
- In case of electric boosting: pre-heating of batch will reduce the need of electric boosting and reduces required electric power use.
- Applicable for recuperative & regenerative furnaces.
- Dry batch (this is an important issue for all pre-heaters) may lead to increased carry-over of batch dust in the furnace & and consequently extra fouling of the regenerators (but this phenomenon is not dependent on the type of pre-heater). The fine fraction in the batch is still present in the preheated material (not entrained in the flue gases) of the indirect preheating system, and may be prone to some carry-over.

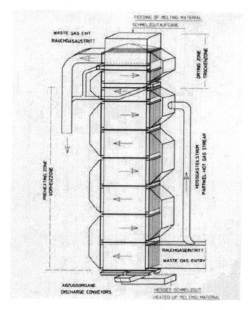

Figure 7   The Zippe indirect Batch & Cullet preheating system

- Due to external cullet with organic contamination: fumes and odour can be released from the cullet in the specially designed top sections of pre-heater. These fumes and water vapour (steam) are withdrawn and are conducted to the main flue gas stream or are incinerated. In one case, an incinerator (Dongen NL) had to be applied to post-combust organic vapours released in first heating-drying zones. Incinerator costs were 350.000 $ (US) plus about 20 Nm^3 natural gas/hr (1-1.5 % of the furnace energy consumption) are required for incineration. It is aimed, in the next generation Zippe pre-heater, that these problems are solved by few design changes.
- Maintenance issues (regenerator fouling and pre-heater cleaning) are reported, requiring extra manpower: about 0.5-0.75 man year/year. However, it is expected that maintenance efforts will decrease in the next generator pre-heater systems.
- In one case, after more than 10 years of operation with a batch preheating system, an extra major repair (> 1 million EURO) of the regenerators appeared to be necessary.

Figure 8 shows the specific energy consumption of an end-port fired regenerative container glass furnace (about 70 % cullet, almost constant cullet ratio: 65-73 %) with a batch plus cullet preheating system, operating since more than 10 years in Europe. During maintenance of the pre-heater (no preheated batch into furnace) the specific energy consumption was about 4950 MJ/metric ton molten glass. After preheating, the pull could be increased with 9 %, and the specific (primary) energy consumption decreased down to 4100 MJ/ton molten glass: a decrease of 17 %.

Since 2006/2007 the German company, Zippe started to develop the next generation Zippe batch and cullet pre-heater system for glass furnaces with the aim of solving the problem of plugging and odour.

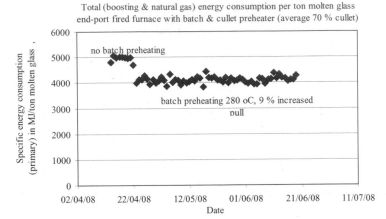

Figure 8   Specific energy consumption of an end-port fired regenerative furnace with electric boosting. The energy consumption is expressed in primary energy
* use per metric ton molten glass.
*The energy efficiency for electricity production in a fossil fuel fired power plant has been taken into account.

Investment cost of typical batch & cullet preheat systems for container glass furnaces:
Typical costs for a complete batch & cullet preheating system, including installation, flue gas bypass and connections, civil works, adaptations of batch charging system of the glass furnace for 300-400 tons glass melt capacity furnaces is 2 million to 3 million EURO excluding filter systems.

A fluid bed batch preheating system has been investigated on industrial scale in the USA in the eighties of the twentieth century by the Gas Research Institute (GRI). The flue gases from the glass furnaces were used to fluidize and preheat the normal batch prior to charging this batch to the glass furnace. Normal batch for container glass production has been heated up to 300-350 °C [16]. Intense dust formation in the fluidized bed and carry-over of the dry batch in the furnace appeared to be problematic as was the economics with the energy costs at that time. The system has been taken out of operation because of the carry-over problems.

New development, the BCP system
Recently PRAXAIR developed a new system that is in the stage of pilot scale experiments and tests. The system is suitable for batch plus cullet preheating (BCP) and dedicated to the use of the flue gases from all oxygen-fired glass furnaces [1]. The flue gases from the furnace are not or only slightly cooled and enter the pre-heater at a temperature level of 1200-1400 °C.
Some features:
- The batch plus cullet (internal and post-consumer) is to be preheated to 900-1100 °F (about 480-600 °C).
- The flue gases from the oxygen-natural gas or oxygen-fuel oil fired furnace are not or hardly diluted with cold air.

- The organic compounds in the cullet are completely removed (fumes are burnt).
- The unit is a modular system and can be adapted to the required size.
- Batch and flue gases can be by-passed in case of maintenance.
- Simple, low-cost system to reduce payback times.
- Special attention to avoid batch-carry over in the furnace, by adapting batch chargers.
- Indirect heat transfer from flue gas to the batch and cullet in the pre-heater.
- Applicable for almost all cullet-normal batch ratios.
- No batch carbon loss (amber glass) by special pre-heater design.
- First stage: indirect radiation heat transfer section, flue gas enters this section without dilution, typically at 1200-1400 °C.

The reported (expected) benefits:

- Fuel savings:	±1 GJ/metric ton glass melt (over oxy-fuel baseline)
- Oxygen savings :	±1/12 ton O_2/ton molten glass
- Production rate increase :	10-20 %
- Emissions Reduction:	15-30 % (CO_2, NOx, particulates) compared to oxy-fired furnace without pre-heater.
- Anticipated payback	between 1-3 years for equipment

A pilot scale system (15 tons batch per day) has been tested by PRAXAIR in 2007 (Tonawanda NY), with preheating temperatures of 480-535 °C. The next step in this development will be the design of a larger system to be applied to a commercial glass furnace.

Important features of batch & cullet preheating systems
The major goal of batch & cullet preheating is to save energy often in combination with an increase of the melt pull (up to 15 %) and in case of electric boosting to save electricity (lower electric currents through the electrodes). The investment costs depend on the size and type of pre-heater and the steel prices, but also on the costs to install the equipment in a glass plant and to connect the flue gas system to the pre-heater. Changes to the batch transport system, often requiring also a by-pass connection and adaptations of the doghouse will add to the capital costs. Pay back times of 3-4 years have been reported, but as energy price levels increase such payback times may decrease in future.

Important aspects associated to batch and cullet preheating are:

- The preheated batch plus cullet is very dry and subject to carry-over during transport and charging of the pre-heated batch plus cullet into the furnace. When, the charged batch is directly exposed to the combustion gases in the combustion chamber of the glass furnace, carry-over of fine-dry batch particles or glass powder may take place. These fine dust particles in the flue gas may react with the refractory materials in the flue gas system: the burner ports and checkers of the regenerators. Especially in the case of end-port fired furnaces, the batch is charged in the direct vicinity of the burners and carry-over may raise a problem of regenerator fouling. (It is recommended to adapt the doghouse design in order to avoid carry-over problems associated with charging of dry batches into the furnace). Figure 9 shows dust concentrations and dust compositions that have been measured in the flue gases of an end-port fired furnace at different positions. The first bar shows the dust content & dust composition just above the checkers, sampling the flue gases from the top part of the regenerator without the pre-heater in operation. The second bar shows the dust contents in the chimney. The difference between the first and second bar presents the amount of dust that has been captured (deposited) in the checkers.

The third bar shows the dust contents in the flue gases sampled from the top of the regenerator (above the checkers) and dust composition, when the pre-heater is operating for a batch of the usual raw materials plus cullet. The dry preheated batch & cullet causes very high dust contents and the difference between the third bar and fourth bar shows that a very large amount of this entrained dust (glass dust, raw materials such as sand, dolomite, limestone, soda) will be deposited in the checkers, resulting in plugging of the regenerators, requiring frequent cleaning or even frequent repairs (e.g. MgO (magnesite) regenerator checker refractory materials exposed to fine sand will show forsterite $MgSiO_4$ formation). A change to coarse raw materials may help to reduce the problem as the last two bars in figure 9 show (batch with coarse raw materials, only 3 % of the batch consists of grains with diameters < 0.1 mm).

The doghouse designs for furnaces, especially for end-port fired furnaces, has probably to be modified to avoid problems as shown in figure 9. It is also important to avoid direct contact between the flames and the freshly charged batch and this may require some modifications to burner positions and burner port designs. For indirect batch & cullet preheating systems, carry-over in the furnace may be more noticeable than for direct pre-heaters, where part of the very fine batch dust is already carried by the flue gases to the filter system.

- For oxygen-fired glass furnaces, the dry batch dusting problem can be controlled by furnace design modifications and proper placement of the oxy-fuel burners, plus choosing the best location for the flue ports, to avoid high gas velocities above the freshly charged batch. TNO conducted dust measurements in the flue gases (flue ports) of two oxy-fuel fired glass furnaces, one with normal wet batch/cullet (not preheated) and one with dry pre-heated batch/cullet mixture. Very little difference in dust contents, comparing the flue gases of both cases, was observed.

- The fumes from external cullet often contain organic vapours or decomposition products of organic components that enter the pre-heater with the post-consumer cullet (residue of food, paper, plastics). These fumes may cause odour and smell problems in the factory or glass plant neighbourhood, dependent on the height of the chimney and level of organic contamination in the cullet. In one case, a post-incineration system had to be applied to post-combust these fumes (conversion into mainly CO_2 and water vapour).

- Preheating of cullet will reduce the amounts of organic materials in the cullet or may almost completely eliminate organic substances by combustion or pyrolysis of the organic material (unless provisions have been undertaken to avoid combustion of carbon in the batch). Thus the oxidation state of the batch will become more oxidized and will fluctuate much less compared to cold non-preheated batches containing post-consumer cullet. Thus, colour and redox control even when using large amounts of fresh post-consumer cullet becomes much more stable. In case of amber glass batches some of the organic material or cokes added to the batch may be lost especially in direct contact pre-heaters.

- Batch & cullet preheating becomes more and more attractive as energy prices will further increase. Batch & cullet preheating presents also a solution when increased production capacities are needed and existing furnaces operate already at their maximum melting rate.

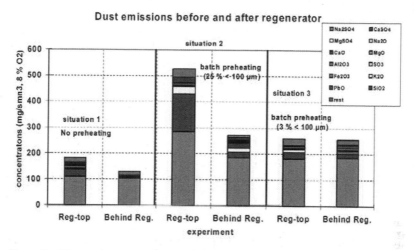

Figure 9   Measured concentrations of dust in exhaust gases of an end-port fired furnace with and without operation of the indirect batch & cullet pre-heater. The compositions of the dust are also indicated. (Reg= regenerator)

CONCLUSIONS

The application of batch and cullet pre-heating is still very limited and centralized in Germany. Depending on the batch – cullet ratios and also emission limit values for certain pollutants (dust, SOx, HCl, HF), one of the here presented systems might be the optimum choice. Batch and cullet pre-heating is probably one of the very few available (BAT) technologies that will result in such high energy saving potential (15-25 %) for glass melting furnaces. Especially for recuperative and oxygen-fired furnaces, the relatively high batch preheat temperatures will offer a large energy saving potential.

Problems of carry-over in the furnace, odour of fumes can probably be solved and technical solutions may exist, but this requires the combined expertise of glass technologists, furnace designers and batch pre-heater engineers.

ACKNOWLEDGEMENT

The author wants to thank to following persons for their contributions and permissions to present the information in this paper:

- Prof. Dr. Hans-Jürgen Barklage-Hilgefort, ARDAGH GLASS, Sven-Roger Kahl, ARDAGH GLASS, Sjoerd Stelwagen, ARDAGH GLASS.
- Dr. Ernst Beutin, INTERPROJEKT.
- Dr. Matthias Lindig, SORG.
- Dr. Sho Kobayashi, PRAXAIR.
- Guenther Mlynar and Dr. Bernd-Holger Zippe, both from ZIPPE Industrieanlagen GmbH.

LITERATURE REFERENCES
1. Kobayashi, H; Evenson, E.; Xue, Y.: *Development of an Advanced Batch/Cullet Preheater for Oxy-Fuel Fired Glass Furnaces.* Proceedings 68th Conference on Glass Problems, 16.-17. October 2007, Columbus Ohio, Charles H. Drummond,   III,   Editor. The Amf . Ceram. Soc. Wiley Interscience (2008) pp. 137-148
2. Beerkens, R.G.C.; Limpt, van H.A.C.; Jacobs, G.: *Energy efficiency benchmarking of glass furnaces.* Glass Sci. Technol. **77** (2004) no. 2, pp. 47-57
3. Herzog, J.; Settimo, R.J.: *Cullet Preheating: The realistic solution for all glass furnaces with cullet argddition.* Ceram. Eng. Sci. Proc. 13 (1992) [3-4] , pp. 82-90
4. Ehrig, R.; Wiegand, J.; Neubauer, E.: *Five years of operational experience with the SORG LONOx® Melter.* Glass Sci. Technol. (Glastech. Ber.) 68 (1995), no. 2, pp. 73-78
5. Ehrig, R.: *Betriebserfahrungen mit LoNOx-Meltern der ersten und zweiten Generation* – Presentation Technical Committee VI of the German Glass Society, Vortrag v. d. Fachausschuß VI d. DGG (Umweltschutz) 15. October 1997 in Würzburg, Germany
6. Kobayashi, H.; Evenson, E.; Miclo, E.: Development of an Advanced Batch/ Cullet Preheater for Oxy-Fuel Fired Glass Furnaces. Glass Trend Workshop on Energy Efficiency & Environmental Aspects of Industrial Glass Melting 10. & 11. May 2007, Versailles, France
7. Schroeder, R.W., Snyder, S.J.; Steigman, F.N.: Cullet Preheating and Particulate Filtering for Oxy Fuel Glass Furnaces. Novem Workshop on Energy Efficiency in Glass Industry. Amsterdam NL, 18.-19. May 2000 Ed. Nohlmans, Moonen, Beerkens, pp. 79-84
8. Barrickman, L; Leone, P.: *Leone Industries: Experience with Cullet Filter. /Preheater.* Proceedings 67th Conference on Glass Problems. 31. October-1. November 2006, Columbus OH, USA Am. Ceram. Soc. Wiley Interscience (2007) pp. 117- 125
9. Barklage-Hilgefort, H.: *Batch preheating on glass melting furnaces.* Glass Trend Workshop on Energy Efficiency & Environmental Aspects of Industrial Glass Melting 10. & 11. May 2007, Versailles, France
10. Enninga, G.; Dytrich, K.; Barklage-Hilgefort, H.: *Practical experience with raw material preheating on glass melting furnaces.* Proceedings of the European Seminar on Improved Technologies for the Rational Use of Energy in the Glass Industry. THERMIE 4.-6. February 1992, Wiesbaden, Germany pp. 129-140
11. Enninga, G.; Dytrich, K.; Barklage-Hilgefort, H.: *Practical experience with raw-material preheating on glass melting furnaces.* Glass Tech. Ber. **65** (1992) nr. 7. pp.186-191
12. Barklage, H.: *Batch Preheating.* 69th Conference on Glass Problems, November 4-5 (2008), Fawcett Center for Tomorrow, The Ohio State University Columbus OH USA
13. Günther Lubitz; Ernst Beutin, Jürgen Leimkühler: *Oxy-fuel fired furnace in combination with batch and cullet preheating.* Novem Workshop on Energy Efficiency in Glass Industry. Amsterdam NL, 18.-19. May 2000 Ed. Nohlmans, Moonen, Beerkens, pp. 69-78
14. Zippe, B.H.: *Reliable batch and cullet preheater for glass furnaces.* Glass Technol. 35 (1994), no. 2, pp. 58-60
15. PLM report (1997) F. Brouwer: *Gemengvoorverwarmer glassmeltoven 16* (Batch pre-heater glass furnace 16) in Dongen Netherlands, final report for NOVEM

16. De Saro, R.; Doyle, E.: Glass Batch Preheater Program. Final report (1987) prepared for the Gas Research Institute, Chicago Ill. USA, report no. 5083-231-0916

INDUSTRIAL RESULTS OF ALGLASS FH OXY FUEL FOREHEARTH BURNER OPERATION

Robert Kalcevic [1], Remi Tsiava [1], Ryuji Fujinuma [1], Chendhil Periasamy [2], Rajeev Prabhakar [2], George Todd [3]

[1] AIR LIQUIDE Centre de Recherche Claude Delorme
1, Chemin de la Porte des Loges – Les Loges-en-Josas – BP126
78354 JOUY-EN-JOSAS Cedex
FRANCE

[2] AIR LIQUIDE Delaware Research and Technology Center
200 GBC Drive, Newark DE
19702-2462
UNITED STATES OF AMERICA

[3] AIR LIQUIDE American Combustion Inc.
200 Chastain Center Blvd, Suite 295
KENNESAW 30144 – GA
UNITED STATES OF AMERICA

ABSTRACT

This paper presents the results of some of the first industrial references for the ALGLASS ForeHearth (FH) burner, an oxy combustion technology developed by AIR LIQUIDE for glass forehearth processes. The ALGLASS FH burner is based on an innovative and patented method for fuel injection with a swirl effect to control flame length. The burner type can also be easily adapted to customer refractory block. The robustness and reliability of the burner and its performances to control flame length and heat transfer were first demonstrated in an AIR LIQUIDE pilot furnace for several months. These advantages have been confirmed through industrial references such as a borosilicate container glass process and an E glass fiber process.

INTRODUCTION

The temperature of the glass which goes out of the melting furnace must be controlled in such a way that the glass obtains adequate properties for downstream operations like forming. When molten glass is flowing in a channel, also called Forehearth or Feeder, a source of heat is generally required to maintain and homogenize the temperature and viscosity of the glass. The heating systems commonly used in glass channels are burners, and the most common fuel is natural gas. Burners used in channels are generally positioned inside a block made of refractory materials. To increase the energy efficiency of burners, air used as oxidizer can be replaced by oxygen-enriched air or preferably pure oxygen. Changing the oxidizer modifies the characteristics of the combustion and more importantly those of the flame. When the oxidizer has higher concentration of oxygen than air, flame temperature increases and radiative heat transfer becomes more intense. Further, the radiative heat transfer occurs in a range of wavelengths which is of more beneficial to glass, contributing to an improvement of heat transfer between the flame and the glass. However, a higher temperature flame can also increase the temperature inside the block, if the flame characteristics are not suitably controlled, which can lead to block overheating/melting.

In comparison with air combustion, oxy combustion flames are shorter when fired at the same power; therefore, there is less concern of flame length when installing face-to-face burners. Air flames

are generally quite long and so that the hot spot of these flames can be located further away from the furnace walls, closer to the center of the chamber. It could lead to two main issues with respect to heat transfer from these flames. The first issue is concerned with the interaction between two opposite flames when the width of the channel is relatively small; the two flames could impinge on each other and be diverted towards the crown of the forehearth channel, causing overheating and degradation of the crown. The flames could also shoot into the furnace chimney, thus causing significant heat losses. Additionally, a longer flame transfers more energy towards the middle zone of the channel rather than where it is really needed, the zone closer to the walls; consequently it could lead to glass reboiling and degrade significantly the quality of the glass.

For efficient utilization of combustion in glass conditioning processes while controlling flame characteristics and optimizing heat transfer, AIR LIQUIDE has developed and patented a technology named ALGLASS ForeHearth (FH) burner.

THE ALGLASS FH TECHNOLOGY

The ALGLASS FH burner design is based on a pipe-in-pipe technology that uses an innovative and patented method for injection of fuel with a swirl effect to control flame length. Oxygen injection surrounds the fuel injection. See the design in Figure 1. Three ranges were developed to cover typical needs in these processes: 0.7-2 kW (2.39-6.82 MBTU/hr), 2-6 kW (6.82-20.47 MBTU/hr) and 6-18 kW (20.47-61.42 MBTU/hr), and every burner type can be adapted to refractory block geometry.

The benefits provided by ALGLASS FH burner were first demonstrated in an AIR LIQUIDE pilot furnace. It was shown that flame structure and length can be controlled.

The burner design helps to control and set the location of the hottest area (or hot spot) of the flame. This parameter is critical to preserve the integrity of both refractory block and burner injector. If the mixing between reactants occurs too rapidly, it can lead to shorter flames and localization of the hot spot within the refractory block. This configuration can cause overheating/melting of the block and/or to the degradation of the burner injector due to formation of soot (or carbon) deposit on the burner injector. Increasing formation of soot gradually changes the flame shape, making the flame shorter and/or less centered within the hole of the block. This can cause impingement of the flame on the burner block and result in damage to the burner block. On the other hand, when the mixing between fuel and oxygen occurs too slowly, the flame is not robust enough and there is the possibility of inadequate mixing of fuel and oxidizer, which can lead to soot formation on the burner injector.

The swirl effect helps to maintain the hot spot of the flame at the same location even when the power of the burner is varied within a wide range. For example when the power increases, the swirl effect on the fuel injection also increases and it increases mixing between the reactants, thus avoiding change in the flame length and hence in the location of the hot spot.

Selection of suitable swirl effect is important to obtain efficient combustion characteristics. If the swirl effect is too strong, the reactants can mix very quickly and produce a shorter flame within the refractory block, resulting in overheating/melting of the refractory block or the burner injector. On the other hand, if the swirl effect is too weak, it does not have sufficient influence on reactant mixing leading to lack of control over the flame length and localization of the hot spot. (Figure 2 illustrates the principle of the swirl effect on fuel injection).

Another benefit of this ALGLASS FH design is its capability to improve and optimize heat transfer to the glass. The flame hot spot is localized outside the refractory block by design, but not too far from the block outlet. Consequently, the maximum heat is transferred to the glass close to the forehearth walls to compensate for heat losses through the walls of the forehearth line and to improve the homogeneity of the glass.

PROCEDURE

In order to prepare for the industrial trials, the refractory block corresponding to the industrial forehearth lines were installed and tested with ALGLASS FH burners in the AIR LIQUIDE pilot furnace without glass. Several diagnostic measurements were employed during the operation of the ALGLASS FH burner to measure:
- The temperature profile in the block, in order to determine the position of the hot spot.
- The temperature on the burner injector in the area where reactants exit to evaluate the risk of soot formation or overheating of the injector.
- The temperature of the refractory block around the outlet hole to assess the homogeneity of the heat transferred all around the block and to verify that there is no damage to the front face of the block (front view).
- The temperature on the chamber bottom to verify that industrial FH line glass temperature is achieved and to appreciate the temperature profile.
- The pressure drop for both natural gas and oxygen inlets.

As a result, the adequate range and the recommended power of the ALGLASS FH burner were selected in order to prevent the risk of damage to both the refractory block and the burner injector and to improve the heat transfer to the glass.

These experimental observations of ALGLASS FH technology have also been modeled with ATHENA, the CFD software developed by AIR LIQUIDE. Turbulence, combustion and radiation models included in ATHENA software have been validated for swirled oxy combustion through comparison of measured and calculated data.

Conversion of one zone in a borosilicate container glass process

The request of the customer, a glass manufacturer specialized in the production of container glass for cosmetic industry, consisted mainly of improving the quality of the glass. Indeed the customer met difficulties to control the very low thickness of certain ranges of bottles. Using air combustion, the customer frequently faced reboil issues in the forehearth line, particularly in the zone downstream the melter exit where the glass temperature is close to 1,300 °C (2,372 F). Another objective of the oxy conversion was, to a leaner extent, to investigate fuel savings.

For the preparation of the oxy conversion, CFD calculations were performed with ATHENA software to choose the best conditions for the trials: predicting the effect of several parameters such as power and positions of burners and assessing the effect of oxy flames on forehearth operation. Before achieving these oxy boosting simulations, the air firing configuration was simulated and validated by experimental data to get a baseline case. The main results of simulations relating to the ALGLASS FH technology showed that a global power close to 50-53 kW (170.61-180-84 MBTU/hr) was suitable for oxy combustion. See Figures 3 and 4 for results of numerical simulations for both air and oxygen cases. Figure 3 shows that the interaction of heat transfer between two face-to-face flames:
- Would occur in the case of air burners (top side of the figure); generating a temperature close to 1,700 °C (3,092 F) near the crown of the forehearth.
- Would not occur using ALGLASS FH burners in these conditions (bottom side of the figure); the temperature near the crown would remain close to an acceptable level of temperature: 1,500 °C (2,732 F).

Figure 4 presents one part of the zone downstream of the melter exit; the impingement between flames:
- Would occur with air burners (left side of the figure); the temperature of the glass surface does not appear uniform along the width of the chamber and the temperature reaches 1,600 °C (2,912 F) in the middle of the channel, which represents a significant risk of glass reboil.

- Would not occur with ALGLASS FH burners (right side of the figure); the temperature of the glass surface appears homogeneous in the area close to the channel walls; moreover the maximal temperature reached in the middle zone is not higher than 1,400 °C (2,552 F), preventing the glass of reboiling phenomenon.

Moreover the simulations showed that the fuel savings would be at least 60 % when compared to the air combustion case.

In parallel, one spare refractory block from the customer was installed and tested with an ALGLASS FH burner in the AIR LIQUIDE pilot furnace. The ALGLASS FH technology was tested at the required power inside the refractory block in order to verify the relevance of:

- The temperature profile measured on the top side of the block.
- The temperature measured in the injection area.
- The temperature measurements around the block outlet hole.

Some of the results of the pilot tests are presented in Figures 5 and 6. Figure 5 shows temperature profiles measured on the top side of the block (blue color) and on the chamber bottom (yellow color):

- The temperature profile of the block shows that the hottest level, around 1,325 °C (2,417 F), is positioned outside of the block.
- The temperature profile on the chamber bottom shows that the temperature remains quite uniform and very close to 1,300 °C (2,372 F) near the block outlet and up to a distance equivalent to the half-width of the industrial forehearth line; indeed, the energy is homogenously transferred towards the bottom of the chamber.

Figure 6 shows measurements of the temperature of the block around the outlet hole of the block. When the chamber bottom of the pilot furnace is about 1,300 °C (2,372 F), the level of temperature measured on the outlet surface of the block is only 20-25 K higher; moreover, the temperature on that surface is pretty homogeneous (+ 6K from bottom to top side), demonstrating the correct centering of the flame inside the block hole.

Industrial tests in a full oxy firing E glass for fiber process

In this industrial forehearth line, the customer already uses oxy burners. The customer was interested in assessing the potential higher heat transfer to the glass of the ALGLASS FH technology and therefore the possibility of decreasing fuel consumption in comparison with existing burners.

To prepare the trials in the chosen zone of the conditioning channel, where the temperature of the glass is close to 1,250 °C (2,282 F), a spare refractory block of the customer was installed and tested in the AIR LIQUIDE pilot furnace using the ALGLASS FH technology. The required power was known; therefore, the burner was tested in this range to appreciate the differences among other things in terms of temperature profiles. See Figures 7 and 8 for pilot furnace results.

Figure 7 shows the temperature profiles measured on the block top for three different powers. Although the power of the ALGLASS FH burner increases, the maximum temperature remains below 1,300 °C (2,372 F) and the profile remains similar. Figure 8 presents the profiles of temperature measured on the bottom of the chamber of the pilot furnace. The temperature close to the block outlet remains close to 1,250 °C (2,282 F) - demonstrating that the location of the hot spot remains fixed. Moreover, when the power is increased, a more homogeneous temperature profile is obtained up to a distance representing the middle of the channel, due to better compensation of heat losses of the pilot furnace.

RESULTS

Conversion of one zone in a borosilicate container glass process

ALGLASS FH burners were installed face-to-face in the 400 mm (15.75") approximately wide zone downstream of the melter exit. Figure 9 shows a schematic top view of the first part of the forehearth line; the oxy combustion zone is positioned close to the air combustion zones. The objective was to bring energy to the glass exiting the melter to improve the homogeneity of the glass while avoiding the reboiling of glass.

During the progressive increase of the power of ALGLASS FH burners to follow the heating-up of the forehearth channel, pyrometric measurements were regularly performed through peep holes to measure the temperature of the burner injector and the temperature of the block around the outlet hole in order to control on one hand the risk of soot formation or block damage, and on the other hand the centering of the flame inside the block. The oxygen/natural gas ratio was higher than stoichiometric conditions to keep the same conditions as in the air combustion zones and to satisfy the customer's request to maintain an oxidizing atmosphere inside the forehearth.

In contrast to the air combustion case, there was no reboil of glass using the oxy-combustion. Moreover this conversion had a significant improvement of the quality and of the homogeneity of the glass at the gob; the control of small thicknesses was improved, what allowed decreasing the rate of rejects. Although, at the time of oxy-conversion, the customer was mainly interested in reducing the rate of rejected material because of the higher cost savings compared to fuel savings, the improvement of heat transfer resulted in 65% reduction in fuel consumption in that zone in comparison with the air combustion case. The ALGLASS FH burners have been using in this zone of the forehearth for more than three years. Figure 10 presents the picture of one ALGLASS FH burner operated outside the forehearth line during an inspection.

Today, due to increasing fuel costs, the fuel savings from oxy-conversion are of greater significance to the customer. As a result, the customer is considering the conversion to oxy combustion of all zones in the forehearth line in 2009.

Industrial tests in a full oxy firing E glass for fiber process

ALGLASS FH burners were implemented in an 850 mm (33.46") roughly wide zone of the conditioning channel where 90 kW (307.09 MBTU/hr) is generally required, representing about twenty four oxy burners.

The burners were ignited and used with an oxygen/natural gas ratio very close to stoichiometric conditions. The flame was checked through peep holes; it remained well centered inside the block. Figure 11 shows pictures of one ALGLASS FH burner: on the left side, the burner is operated inside the forehearth line, the well centered flame inside the refractory block can be noticed; on the right side, the burner was pulled out for inspection after several weeks of continuous operation; there was no sign of burner damage or glowing injector. Over six months of operation, the flames remain showing good shape and there has been no damage to the burner blocks. There has been minor random deposits looking like glass noted in the lower side of the burner body. Further, due to pull rate variation, the power of the burner was changed by 50%; this caused some damage to the burners. Trial is still ongoing and actions are in progress to resolve the issue.

After confirming the robustness and reliability of the ALGLASS FH technology over a longer period, by avoiding the random glass deposits, the next step shall consist in assessing potential energy savings in comparison with existing oxygen burners.

CONCLUSIONS

For the borosilicate container glass process: preparation through modeling calculations, pilot tests using the ALGLASS FH technology and a spare customer refractory block allowed to set and validate the required power for oxy combustion in the zone downstream the melter exit and to estimate the fuel savings due to oxy conversion. ALGLASS FH burners using natural gas and oxygen have been installed face-to-face replacing air burners and using the same refractory blocks. In contrast to air combustion case, there was no reboiling of glass. Moreover this conversion significantly improved the quality and the homogeneity of the glass at the gob, improved the control of small thicknesses, and reduced the amount of rejected material. The improvement in heat transfer reduced fuel consumption in that zone by around 65 % in comparison with the air combustion case, due to localization of the most energetic zone of the flame outside of the block.

ALGLASS FH burners have been using in this zone of the forehearth for more than three years. Today, due to rising energy costs, the customer is considering the conversion of all zones in the forehearth line to oxy combustion in 2009.

For the E glass fiber process: at first, tests in the AIR LIQUIDE pilot furnace were performed with the ALGLASS FH technology using a spare customer refractory block in order to control the temperature profile within the block and validate the efficient utilization of the ALGLASS FH technology for the refractory block and for the burner itself. Several existing oxygen burners were replaced by ALGLASS FH burners in one zone of the conditioning channel. The flames remain showing good shape; there has been no damage to the burner blocks. However, there has been minor random deposits looking like glass noted in the lower side of the burner body. Further, due to pull rate variation, the power of the burner was changed by 50%; this caused some damage to the burners. Trial is still ongoing and actions are in progress to resolve the issue.

After confirming the robustness and reliability of the ALGLASS FH technology over a longer period, by avoiding the random glass deposits, the next step shall consist in assessing potential energy savings in comparison with existing oxygen burners.

Fuel

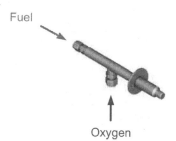

Oxygen

Figure 1: scheme representing the ALGLASS FH pipe in pipe burner with fuel and oxygen inlets

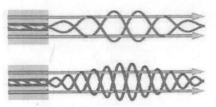

Figure 2: Scheme symbolizing the fuel (red color) and the oxygen (blue color) flows when the power changes; thanks to the swirl effect, the reaction between reactants takes place in the same area which leads to the same flame length

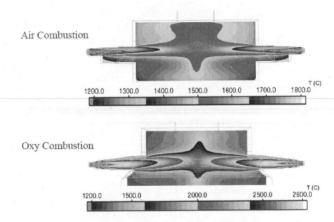

Figure 3: Modeling of two face-to-face burners in the zone downstream of the melter exit – for borosilicate container glass process: air combustion case (top) and oxy combustion case (bottom)

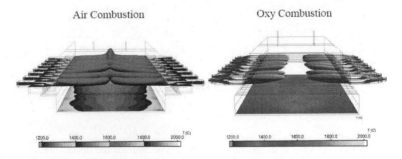

Figure 4: Modeling of face-to-face burners in one part of the zone downstream of the melter exit – for borosilicate container glass process: air combustion case (left side) and oxy combustion case (right side)

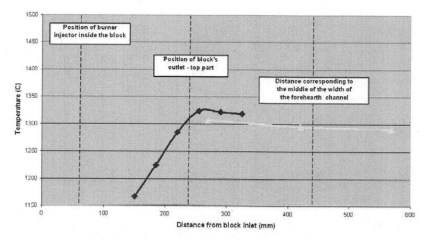

Figure 5: Graphic – Tests in the AIR LIQUIDE pilot tests – Temperature measurements on block top (blue color) and on chamber bottom (yellow color) when the ALGLASS FH burner is operated inside the customer refractory block – for borosilicate container glass process

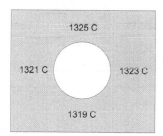

Figure 6: Scheme front view – Tests in the AIR LIQUIDE pilot tests – measurements of the temperature of the block around the outlet hole of the block when the ALGLASS FH burner is operated inside the customer block – for borosilicate container glass process

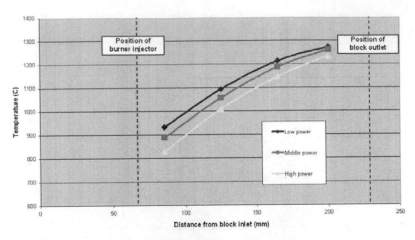

Figure 7: Graphic – Tests in the AIR LIQUIDE pilot tests – Temperature profile on block top when the ALGLASS FH burner was operated at three different powers inside the customer's refractory block – for E glass fiber process

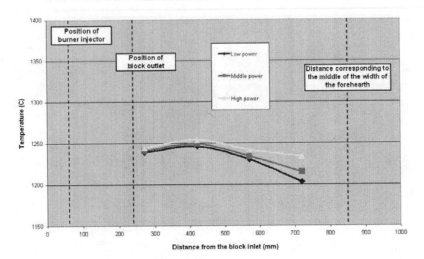

Figure 8: Graphic – Tests in the AIR LIQUIDE pilot tests – Temperature profile on chamber bottom when the ALGLASS FH burner was operated at three different powers inside the customer's refractory block – for E glass fiber process

Glass flow

Figure 9: Schematic top view of the first part of the borosilicate container glass forehearth line where oxy combustion zone, downstream of the melter exit, is represented (left side)

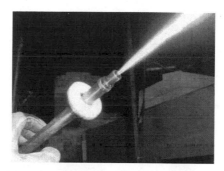

Figure 10: Picture of one ALGLASS FH burner operated outside the borosilicate container glass forehearth line

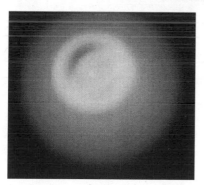

Figure 11: Pictures of one ALGLASS burner: operated inside the E glass fiber forehearth line (left side) and pulled out for inspection after several weeks of operation (right side)

FLOAT FIRE GAS FIRED SYSTEM FOR TIN FLOAT LINES

James B. Roberts
Eclipse Combustion, Rockford, IL.

Herb Gessler
Gessler Engineering, Pittsburgh, PA

Gary Deren
Unifrax Corp., Niagara Falls, NY

ABSTRACT

In the 1950's Sir Alastair Pilkington and Kenneth Bickerstaff of UK's Pilkington Brothers developed a forming process where a ribbon of glass flows on top of a shallow bath of molten tin to form a continuous ribbon of glass. The Float Process can now be considered one of the great glass achievements of the 20th century along with the Corning Ribbon Machine, I-S bottle machine, and the Fiber Forming process, which make up the bulk of today's glass industry.

The Tin Bath, at the heart of the float process, develops the ribbon of glass to the desired thickness and width under a controlled inert atmosphere of nitrogen and hydrogen. An electrical heating system, using silicon carbide electrodes, is used to control the temperatures within the Tin Bath as the glass is pulled and cooled from about 1100°C to 600°C. The electric firing system, while considered a standard component to the Tin Bath, does has several inherent limitations.

In the mid 1990's Gessler Engineering developed and engineered the concept using a radiant tube burner system in the Tin Bath. The Float Fire® system was later tested and proved with over 5 years of continuous operation as an auxiliary heating system. Gessler Engineering along with Eclipse, Inc. and Unifrax Corporation has taken the concept further to develop designs for a completely operational Tin Bath using the Eclipse SER Radiant burner technology. In this paper we intend to show and demonstrate the advantages of the Float Fire® technology.

HISTORY

In January 1999, the United States Department of Energy published a report *Advanced Ceramics in Glass Production: Needs and Opportunities*, which was a direct result of a workshop with participants from the ceramics industry, the glass industry and the government. From this seminar **"Exhibit 9 – Critical Needs in the Float Furnace"** identifies …"gas-fired radiant burners without oxygen, that are muffled in a radiant tube, with low operating costs and turndown capability" as the highest priority for design implementation. Surprisingly, this "think tank" of industry experts unknowingly described and validated what was already conceived and patented by Gessler Engineering several years previously.

The **"Float Fire"** Bath Roof, today, is the end result of an evolutionary process spanning the last seventeen years.

INTRODUCTION

It is common knowledge that heating with natural gas is more energy cost-efficient than heating with electricity. Changing process heating from electricity to gas is certainly not a novel idea. However, particular problems arise during the flat glass production when the conversion is applied to the Tin Float Bath. One such problem encompasses contamination of the bath inert gas atmosphere to the detriment of both the glass and the tin. This presentation provides information for the conversion, the solution of the associated problems and the resulting benefits.

In the Tin Float Bath process of flat glass production, an entirely new operational bath roof is submitted to replace the present roof in use for the past forty years. The new roof comprises two parts, the utilization of a newly designed roof fabrication and the use of automatic recuperative natural gas burners instead of the presently used electrical heating elements. Individually, both the fabrication and the burners contribute to a cost-efficient operation the highly excels the present operation, amounting to a calculated Sixty + percent (60+%) savings.

GENERAL

The fabrication of the new roof housing utilizes a steel plate shell of such a required depth to allow the installation of the gas burners in the vertical position. The interior of the new bath roof essentially consists of a hot face sillaminite block layer, similar to that used presently. Utilizing a ceramic fiber lining behind it, this maintains an approximate temperature differential from 2200 degrees F inside to 175 degrees F outside the shell. This would be near the hot end. No cooling chamber is needed above the new bath enclosure as presently required to protect the existing electrical equipment and heating components of the present Tin Float Bath.

In the new "**Float Fire**" Bath roof, Eclipse Self Recuperative natural gas burners will eliminate the use of the electrical heating elements. Combustion air and natural gas are ignited and burned within a heat-radiating tube. The products of combustion are exhausted externally and have no contact whatsoever with the internal bath inert gas atmosphere.

These gas burners employ no moving parts, and allow for easy insertion of the burners through mounting flanges even during full operation. There are only two piping connections for gas and combustion air. They are automatically spark ignited and can utilize either Flame ionization style flame monitoring, or Ultra violet scanning. The control system is a simple HIGH-FIRE, LOW-FIRE, OFF control, or a modulated control system can be employed as well. The burners are considered low emission burners by present technology standards. Process/Line control of the burner zones is comparable to that utilized with the control of the electrically heated zones.

This new "**Float Fire**" Bath roof surpasses the presently used bath by providing an operation that is more energy cost-efficient. It is more readily controlled with little or no shut downs, by providing an installation that is less labor-intensive, and is easily maintained. These burners, along with the improved roof design is thermally efficient, is anticipated to have a longer life expectancy.

BENEFITS

The use of **Eclipse Auto-recupe** self-recuperative natural gas burners in the tin float bath process will result in the following operational cost benefits. The utilization of natural gas is

more energy cost-efficient than electricity, rendering approximately a sixty (60) percent reduction in fuel/energy dollars.

The new roof also incorporates cost saving design features, such as, bolted gasketed segments, of lesser weight, which are more easily handled during construction or repairs. The burner installation encompasses few moving parts, with easy insertion through the mounting flange even during full operation, which provides for continuous operational capability, unlike present bath roof installations. Further, this results in savings arising from maintenance, scheduled shutdowns, demolition and repairs.

The burner has only two piping connections for gas and combustion air, and a simple control system, proportionately modulated or HIGH-FIRE, LOW-FIRE, OFF. Existing control systems can be readily adapted to the new operation with simple program modifications. Furthermore, the gas burners are utilized at approximately sixty percent (60%) of their rated capacity, which promotes longevity. If a single burner does require replacement, it can be readily removed and replaced from the outside. Each burner can also be isolated from the Zone with simple manual shut-offs, essentially blanking it from the zone until repairs can be effected (Note: The replacement burner can be fired immediately after insertion with no detrimental effects from thermal shock).

Considering the sixty percent (60%) utilization of the rated capacity, the proposed burners could preclude the function of auxiliary burners presently used to preheat the bath prior to glass flow. The gas burners are more readily controlled to evenly heat and dry the internals, obviating the detrimental effects of open-flame heating equipment. In the event of a glass ribbon loss, the burners can be used at a higher heating capacity to bring the Tin Float Bath back to operational temperatures within a shorter time frame, minimizing production losses.

The simplicity of the natural gas heating system with the use of one recuperative burner type throughout the bath contributes to reduced emergency inventory, minimal storage area and a short learning curve for maintenance personnel.

Also, with the elimination of the internal electrical heating system components and the associated cooling chamber, the new roof design affords a substantial reduction in the volume of inert gases used during production. Without the high volume cooling requirement, the amount of inert gases used will be only what is necessary to maintain positive pressure inside the bath to prevent oxygen infiltration.

RELATED COST SAVINGS

The most obvious of the related cost savings are derived from the elimination of the equipment relative to the present electrical bath roof. For new bath roof design installations, electrical equipment such as: substations, transformers, switchgear, bath heat power panels or SCRs, power bus ducts, transformer/reactors, bus tap boxes, bus bars, heating elements, all interconnecting cables and connectors, tray and conduit, etc. will not be required. The new "**Float Fire**" Bath Roof design also affords a substantial reduction in its inert gas atmosphere operational costs. The above lists the most apparent areas for cost reduction. The list is not inclusive. The full extent of the related cost savings can only be determined after the installation and operation of the "**Float Fire**" bath roof.

CONCLUSION

The Tin Float Bath used today has remained essentially the same since its introduction almost 50 years ago. The corrosive atmosphere of the bath process has limited the past improvements to those that can withstand the hostile environment. Electrical heating elements were the only solution to provide the requisite heat. The bath has considerable thermal loss. In addition, the bath roof has a limited life that requires costly periodic reconditioning. In terms of research and development, the process has virtually reached its maturity. Although the existing Tin Float Bath is the standard method for the flat glass production to date, manufactures have had to contend with its negative capital aspects.

The "**Float Fire**" Tin Float Bath Roof was designed to eliminate these inherent problems. The advents of recently developed high temperature materials, along with advances in combustion science, have demonstrated a history of reliability and cost-effectiveness. These developments have made such revolutionary changes in the bath roof construction feasible. The newly designed roof and insulation fabrication, coupled with the proposed alternative heating method, contribute to an energy cost-efficiency and a greater expected longevity that will out perform the conventional bath roof. Theoretically, periodic reconditioning of the new bath is not necessary.

With substantial increases in energy costs and the rising demands on electrical power, industry efforts must be concentrated on cost reduction and efficiency improvements. The slightest improvement to any process can provide a financial advantage for a company, affording it a competitive edge. Energy cost-efficiency and reduced maintenance expenses translate into a more profitable operation. The savings can be considerable, given the vast quantities of glass produced each year in the industry. A potential for improvement exists in the tin float bath process. The "**Float Fire**" Bath Roof is the logical progression for this technology.

REFERENCES

Excerpts from Gessler and Eclipse publications and documents have been referenced and utilized in this presentation.

EVALUATION AND IMPLEMENTATION OF X-RAY ANALYSIS OF RAW MATERIALS USING BORATE FUSION SAMPLES

Neal T. Nichols and Brian D. Mitchell
West Analytical Services

ABSTRACT

West Analytical Services maintains a Materials Testing Laboratory in Maumee Ohio with an emphasis on glass and glass raw material testing. In this lab, glass composition has been determined using Wavelength Dispersive X-ray Fluorescence since the 1960's. Most raw material analysis has been performed using classic wet chemical and spectroscopic techniques. In order to better respond to customer timelines we are now performing analysis of selected elements in sand, feldspar, dolomite and limestone with WDXRF following fusion of the sample with lithium tetraborate. Our approach is to use those X-ray results where we can establish reliability and to use other techniques (e.g. ICP, AA or wet analysis) where detection limits or other factors limit the precision or accuracy. The determination of which elements could successfully be determined by X-ray was made through comparison of X-ray data and wet chemical or standard values. The evaluation process and the results will be discussed in this presentation.

INTRODUCTION

Our lab has been serving the glass industry for over thirty years through testing and analysis of finished product, defect analysis and determination of raw material composition. Finished product is tested for various physical properties and composition. Where standards are available most glass compositions are determined using Wavelength Dispersive X-ray Spectrometry (WDXRF). For compositions where appropriate standards are not available which contain elements not well detected by WDXRF or at levels too low to be detected, wet chemical determinations are employed.

Raw material composition has traditionally been determined using wet chemical methods in our laboratory. Although at various times in the past pressed pellets and borate fusion disks have been analyzed by WDXRF, the problems with each of these techniques has led to a return to wet methods. Pressed pellets of finely ground materials offer acceptable detection limits but the presence of some hard minerals which cannot be easily ground leads to inhomogeneous samples and results. Lithium tetraborate fuses these minerals and gives a homogenous sample, but the dilution of sample with flux (1:8) can increase detection limits for some trace elements to an unacceptable level. Although the wet chemical methods have excellent accuracy and precision, they are labor intensive and require skilled chemists. Due to the substantial amount of time required for wet analysis it is necessary to batch samples for analysis to keep the cost effective. The time required for analysis and holding samples for batching leads to turn around times which can be problematic for customers. Alternative approaches to raw material testing have been considered in an effort to reduce turnaround time:

1. ICP/AA analysis of all elements.
   Material is put into solution(s) with various acids and run spectroscopically. This is the preferred method for many trace and minor components but due to the inherent error of the technique (around 5%) this is generally not acceptable for major components.

2. WDXRF analysis of fused samples for all elements.
   Lithium tetraborate fusion pellets are analyzed for all elements. This method may be adequate when appropriate standards are available but detection limits may be too high for some trace elements due to the dilution of sample by flux.

3. Use of WDXRF analysis of lithium tetraborate pellets for some elements and ICP or AA analysis of sample solutions for elements where WDXRF is not adequate.

Our current approach to raw material analysis is using the combination of WDXRF of fusion pellets for major components and other elements where we have been able to demonstrate good correlation with wet results. For those elements where detection limits or inadequate standards do not allow good correlation, we will continue to use wet chemical methods. The determination of the appropriate technique for each element is based on comparison of wet and WDXRF analyses of samples and standard materials. At this point, we have limited this transition to sand, feldspar, dolomite and limestone.

The first step in this change was acquiring an automated fusion apparatus. This was straightforward as there are several types available commercially. We chose one based on specifications and user recommendations. Another option is to prepare the pellets without an automated instrument and the literature has many examples of homemade systems. As the time required for preparation and consistency of samples were important for our needs, we chose to purchase an automated unit which has met our expectations.

In WDXRF, the analysis is limited by the quality of the standards used. We have all the applicable available NIST SRM's and many standard materials which were analyzed in our lab. As we were selecting standard materials for the calibrations, it became clear that there is limited availability of good standards. Unlike a vendor of a raw material who can use a type standard of their material and get adequate results, we receive samples where the composition may vary widely. To handle these samples we need standards to bracket these variations. For sand we currently have six standards and are evaluating three which we just acquired from England. The silica values in sands have had poor correlation with wet results thus far and we hope the incorporation of additional samples will improve this. For feldspar, we are using seven standards and are considering additional standards to improve the aluminum and calcium results. For limestone, we are using five standards with good results. For dolomite we are using three standards with good results, but we may acquire more to allow for compositions which vary further from suite of standards.

ANALYTICAL APPROACH
Traditional Methods:
1. Sand
   Representative sample is ground in agate to a fine powder.
   Silica is determined gravimetrically.
   Al, Fe, Cr, and Ti are determined by ICP.
   Na, Mg, K and Ca are determined by AA.
   Loss on ignition is determined.
2. Feldspar
   Representative sample is ground in agate to a fine powder.
   Silica is determined gravimetrically.
   Aluminum is determined volumetrically.
   Fe, Cr and Ti are determined by ICP.
   Na, Mg, K and Ca are determined by AA.
   Loss on ignition is determined
3. Limestone
   Representative sample is ground in agate to a fine powder.
   Calcium is determined volumetrically.
   Si, Al and Ti are determined by ICP.
   Na, Mg and K are determined by AA.

Fe and Cr are determined by ICP or colorimetrically.
Loss on ignition is determined.

4. Dolomite

Representative sample is ground in agate to a fine powder.
Calcium and Magnesium are determined volumetrically.
Si, Al, and Ti are determined by ICP.
Na and K are determined with AA.
Fe and Cr are determined by ICP or colorimterically.
Loss on ignition is determined.

WDXRF/Borate Fusion Method

Representative sample is ground in agate to a fine powder and dried at 105°C for several hours. 1(+/- 0.0004) gram of sample and 8 (+/- 0.0004) grams of lithium tetraborate are weighed into a 95% Pt, 5% Au fusion crucible. The mixture is fused over a propane oxygen flame on an automated fusion apparatus. The fusion material is poured onto a Pt/Au mold, cooled and transferred to an X-ray spectrometer for analysis against standards prepared in the same manner.

EVALUATION OF DATA

Tolerances have been defined for various analytical techniques employed in our laboratory based on evaluation of results of duplicate analyses, known error associated with techniques and results of analyses of standard materials. Based on these general guidelines we routinely analyze standards along with samples and verify that the results are within the established range as a quality control check. As these tolerances were already in place for the materials of interest it was decided to use the wet lab tolerances to evaluate the X-ray analysis results of the fusion samples.

1. Sand

   SiO_2 (based on gravimetric analysis) +/- 0.2%
   Al_2O_3, Fe_2O_3, TiO_2, CaO, MgO, Na_2O, K_2O and Cr_2O_3 (based on instrumental analysis) 5% relative

2. Feldspar

   SiO_2 (based on gravimetric analysis) +/- 0.2%
   Al2O3 (based on volumetric analysis) +/- 0.2%
   Fe_2O_3, TiO_2, CaO, MgO, Na_2O, K_2O and Cr_2O_3 (based on instrumental analysis) 5% relative

3. Limestone

   Ca (based on volumetric analysis) +/- 0.2%
   SiO_2, Al_2O_3, Fe_2O_3, TiO_2, MgO, Na_2O, K_2O and Cr_2O_3 (based on instrumental analysis) 5% relative

4. Dolomite

   CaO and MgO (based on volumetric analysis) +/- 0.2%
   SiO_2, Al_2O_3, Fe_2O_3, TiO_2, Na_2O, K_2O and Cr_2O_3 (based on instrumental analysis) 5% relative

In order to evaluate the WDXRF/Borate Fusion technique, samples were analyzed using both the traditional wet lab methods and the new technique. Results were compared using the established

tolerances and suitability of the new technique was determined for each element in the four sample types. Since we analyze all elements (none by difference) and determine loss on ignition on samples, we also consider the total when evaluating analysis results. A poor total indicates unanalyzed elements or problems with the analysis. At this time, nineteen sand samples and standards, ten feldspar samples and standards, ten limestone samples and standards and three dolomite samples have been evaluated.

RESULTS

Element (Oxide)	Sand	Feldspar	Limestone	Dolomite
SiO_2	*	Acceptable	Acceptable	**
Al_2O_3	*	**	Acceptable	**
Fe_2O_3	*	Acceptable	Acceptable	**
TiO_2	*	Acceptable	Acceptable	**
CaO	*	**	Acceptable	Acceptable
MgO	*	Acceptable	Acceptable	Acceptable
Na_2O	*	Acceptable	**	**
K_2O	*	Acceptable	Acceptable	**
Cr_2O_3	*	**	**	**

* Sand

Silica results were not within the desired tolerances. The X-ray results were generally around 0.6% lower than wet or standard values. We have purchased additional sand standards and will establish a new calibration to repeat the evaluation. Until then we continue to report wet lab results.

** Correlation of wet chemical and WDXRF/fusion results not adequate. Wet chemical results are reported.

CONCLUSION

In order to respond to our customers need for more rapid analysis of raw materials, our lab evaluated changes in our approach to these analyses. Although classic wet chemical analysis will generally give the best precision and accuracy, the time and technical skill level required for these analyses is often not practical. WDXRF is a reliable rapid technique which we have successfully employed for analysis of glass and have also had limited success with raw materials in the past. Based on our previous experience, we chose not to pursue pressed pellets or powders due the mineral effect. Borate fusion pellets provide homogenous representative samples, but the dilution with flux negatively impacts the detection limits of trace elements. Our approach is to use fusion pellets with WDXRF for major components which take the most time in the wet chemical lab. For minor and trace elements we will use the WDXRF results where we can demonstrate good correlation with traditional wet results. In those cases where good correlation is not seen, we will continue to use wet analysis. By merging these techniques, we hope to decrease analysis turnaround time while maintaining the reliability of results. We consider this approach to be successful as we have already switched many of the time consuming analyses to WDXRF. As we acquire more standards and refine the calibrations to better match the samples we receive, the utility of this technique will increase.

GLASS SURFACE CORROSION AND PROTECTIVE INTERLEAVING SYSTEMS

Paul F. Düffer
Industry Relations Chair
Society of Glass and Ceramic Decorators
PPG Industries, Inc., Retired 2008

ABSTRACT

The historical development of glass interleaving systems is addressed within the context of the phenomenon of glass surface corrosion. Early twentieth century investigations into the viability of using paper as a separation medium in the flat glass industry are explored in order to establish the technological foundation that underpins the subsequent introduction of powdered materials. In addition, contemporary advances in the use of anti-corrosive interleaving materials are summarized using information gleaned from the patent literature as well as promotional brochures and information disseminated by the vendors of such materials. Finally, several examples of corrosive damage to glass will be presented in order to demonstrate the challenges that persist in maintaining surface quality during post-production storage and handling.

INTRODUCTION

The application of interleaving materials in flat glass packaging is a practice that has been undertaken since the advent of crown glass manufacturing centuries ago. Initially, the intended function of interleaving materials was most likely that of making it possible to separate stacked sheets with the additional objective of protecting the glass articles from physical damage during transport. One prominent interleaving material utilized during these early days of flat glass manufacturing was straw. In fact, straw was used as an interleaving material throughout the first decade of the twentieth century when it was replaced with other materials the most prominent of which was paper.

As mentioned, the original function of interleaving materials was to provide for easy separation of flat glass sheets and the prevention of physical damage during transport. It was not until the early twentieth century that surface corrosion became an additional focal point of interest. In 1666 the chemist, Robert Boyle, was attracted to the white residue left behind when he distilled water from a glass vessel. After two hundred iterations, the residue continued to be left behind. It was almost 100 years later in 1770 when Lavoisier carried out experiments that established the residue of Boyle to have emanated from the glass. In a comprehensive review of the durability of glass, Newton [1] describes the relative ignorance that prevailed at the beginning of World War I with respect to the inherent durability of both container and flat glass products. As a consequence, the most useful developments in the field of interleaving technology and the study of glass surface corrosion did not occur until well into the twentieth century.

THE PERIOD FROM 1900 TO 1950

Glass Surface Corrosion – Early Concepts

By the early twentieth century, engineers and technologists in the flat glass business were aware that under certain conditions glass reacts with water in a manner that can have an adverse impact on surface quality. In 1913, one proposed mechanism suggested that glass components such as sodium and calcium silicates were dissolved in minute quantities by the water in contact with the surface. The calcium silicate subsequently becomes hydrolyzed to set forth hydrate of lime and hydrated silica. It was also observed at the time that increasing the soda content causes glass to be less resistant to staining. Upon evaporation of the water, the residues deposit on the glass dimming the surface. When the process persists for an extended period, the glass surface will be destroyed by scales and minute crystals of silica. If this process occurs in the presence of carbon dioxide, sodium carbonate forms in

solution creating an alkaline environment that exerts a greater corrosive action on the glass than water alone. However, in a dynamic environment such as glass in a window frame, the customary and natural cleaning of the glass by the elements will keep the surface in fairly good condition.

When it came to storing glass, the conventional wisdom of the time maintained that any film or dust remaining on the glass surface could trap moisture and result in corrosion. For this reason, plate glass should be thoroughly cleaned before storage in order to remove the remnants of polishing compounds and grouting materials. In addition, it was advised that damp straw should not be used for packing the glass. However, interleaving material selection was not the only focal point of attention in the ongoing effort to control stain. Rather, there was also a concerted effort underway within the industry to develop a stain resistant glass.

Additional insight into the level of awareness regarding glass surface corrosion can be found in a patent issued to E.D. Brogley in 1919 [2]. The objective of the invention was to eliminate what was described as the rotting of straw interleaving in the presence of moisture that resulted in discoloration and staining of the glass. Fumes generated by the rotting straw "burn" the glass to the point where the glass may at times become opaque. The proposed remedy called for treating the glass surface with a mixture containing lubricating oil, sulfuric acid and water followed by packing in straw.

Over the next 30 years, few advances were made in elucidating the mechanism of glass surface corrosion by means of detailed laboratory studies. However, the absence of such information did not dissuade the glass industry from pursuing advances in glass interleaving materials.

Interleaving Developments

By the early 1930's, the major flat glass manufacturers had transitioned from straw to paper interleaving materials since paper was found to be more effective in reducing the incidence of staining on glass. One of the first papers used successfully as an interleaving material was newsprint. However, newsprint was found to leave an unsightly "scum" on the surface when glass was stored in a humid environment. In response to this observation, the flat glass manufacturers of the day initiated a number of studies to identify the ultimate paper product suitable for packaging their products. Among the paper materials tested were ham wrap, tire wrap, rope gasket, tan jute and folder stock as well as kraft paper. By the mid-30's, studies had pointed out that newsprint offered the best protection against corrosion but also produced the most "scum" deposits. On the other hand, the least amount of scum was found when kraft paper was used, but kraft paper was inferior to newsprint in its ability to control surface corrosion. As a consequence, efforts were initiated to identify methods for enhancing the anti-corrosive properties of kraft paper.

Attempts to enhance the anti-corrosive properties of kraft paper included treating the paper with hydrochloric acid, aluminum sulfate, ferric sulfate and copper sulfate. However, these investigations met with little success as the decade of the 30's came to a close and isolation of the "perfect paper" remained an elusive goal. Nonetheless, advantageous properties of a suitable interleaving paper were identified to include: 1) Production from sulfate pulp. 2) A pH less than 5.0. 3) The ability to absorb moisture. 4) A propensity to resist wrinkling. In addition, prominent proponents of interleaving technology at this time viewed the ability pf a paper material to absorb moisture as a major feature in controlling surface corrosion.

From the 1930's until the late 1940's, paper interleaving remained the material of choice for packaging glass. Both newsprint and kraft papers continued to be used by the industry for both sheet and plate glass. In the absence of major improvements in paper interleaving performance, flat glass manufacturers stressed the importance of ventilating storage areas and minimizing the potential for the formation of condensation on glass. In addition, it is worth noting that temperature was a well-known driver of glass surface corrosion.

THE PERIOD FROM 1950 TO 2008

Glass Surface Corrosion Studies – Post WWII Developments

In1949, studies of glass surface corrosion were being undertaken by several prominent university groups. Foremost among these were investigations conducted by Douglas et al at the University of Sheffield [3]. By 1967, these studies established that for commercial flat glass a pH of 9.0 represented a critical point at which glass network dissolution occurred [4] [5]. Furthermore, it was established that moisture-induced glass corrosion proceeds by means of a two-step process in which step one involves a first order ion exchange process that increases solution pH followed by dissolution of the glass network when the pH reaches 9.0. In particular, hydrogen ions (hydronium ions; see Ernsberger [6]) from the contacting water exchange with sodium ions in the glass in a manner that increases the net concentration of hydroxyl ions in the water leading to increased pH levels. When pH reaches 9.0 and above, dissolution of the glass network occurs resulting in what is commonly know as glass surface corrosion or staining. Figure 1 illustrates key features of this process as reported by Düffer in 1993 [7]. The straight-line plot of sodium leaching versus root time under conditions of controlled pH confirms the proposition that this reaction proceeds via a first order ion exchange mechanism. Commonly known as Stage I corrosion, this reaction does little more that result in creation of a hydrated glass surface for which the optical quality remain essentially unaffected. On the other hand, the plot of sodium leaching in distilled water presents a far different result. During the early phase of the reaction a first order mechanism predominates. However, as solution pH begins to increase a marked deviation from the diffusion-controlled process occurs as dissolution of the glass network prevails. This reaction phase is often referenced as Stage II corrosion. Unlike Stage I, Stage II corrosion results in permanent surface damage for which there is no cost-effective means for achieving restoration. An example of such damage is shown in Figure 2.

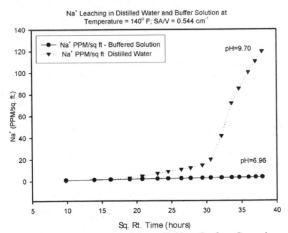

Figure 1. Stage I and Stage II Glass Surface Corrosion

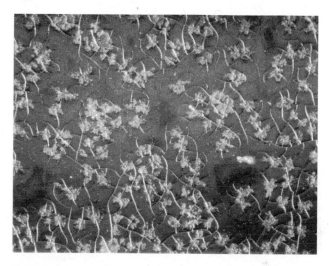

Figure 2. Glass surface damage resulting from Stage II corrosion, 200X

A detailed review of corroded glass surfaces has been published by Clark et al [8] that includes an extensive bibliography which cites a number of investigations into the glass corrosion process.

The marked dependence of Stage I glass surface corrosion on temperature has also been previously reported [7]. Figure 3 shows the affect of temperature on the rate of increase in pH levels for commercial float glass samples maintained in contact with distilled water at 72°, 140° and 195° respectively. In an uncontrolled environment, glass network damage ensues when pH levels approach 9.0 and above. This pronounced affect of temperature accounts for statements from veterans of the industry indicating that surface staining was typically a spring and summer phenomenon; the fall and winter months bringing a welcomed forgiveness from corrosion in temperate regions of the USA.

The positive outcome of the academic studies was to provide clues for the development of functional glass interleaving systems that could control the onset of Stage 2 corrosion or the network dissolution of glass in storage. As a result, even though it may not be possible to keep glass from encountering moisture, interleaving materials could now be selected on the basis of controlling pH in stacks of flat glass sheets.

Glass Interleaving Developments

On the threshold of the 1950's, Gwyne & Williams were awarded a patent for an apparatus to apply wood flour interleaving to flat glass surfaces [9]. Even though paper continued to be used for glass interleaving, there was a distinct shift towards developing powder based technologies that would be functional as well as less costly than paper. From 1950 to 1969, a number of powdered materials were tested for potential use in packaging glass. Harwood and briarwood flours were evaluated along with ground walnut and coconut shells. However, none of these materials were found to be an acceptable replacement for paper.

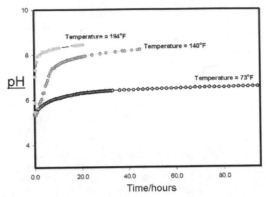

Figure 3. The influence of temperature on the rate of increase in pH during Stage I glass surface corrosion.

By 1969, attention began to focus on the potential use of polymer based interleaving systems. The introduction of float glass production during this period increased demands for a functional interleaving powder that could be applied in a manner capable of accommodating the large & rapid glass output of this process. Hay in 1973 [10] described an interleaving system that consisted of salicylic acid agglomerated with polyethylene oxide and mixed with an inert separator such as polystyrene beads. This combination of materials was reported to be as inexpensive as wood flour to apply but to be more effective in controlling stain. Simpkin & Ascroft [11] followed Hay by developing a powdered interleaving system based upon the combination of wood flour impregnated with adipic acid that is blended with polymethylmethacrylate (PMMA) beads. Albach [12] proposed using ammonium chloride in combination with wood flour as an interleaving material. In particular, the ammonium chloride was applied by spraying an aqueous solution to the glass followed by deposition of dry wood flour to the wet glass. Thus forming an "adherent coating on each glass sheet". Today, most powder interleaving material used to package glass consists of adipic acid blended with PMMA beads that are applied in dry form to the glass.

These representative advancements in powder interleaving technology were followed by the introduction of several supplemental interleaving systems for use in combination with powdered materials. Düffer [13], [14] and Franz [15] introduced technologies that involve the spray application of mixed organic acids and buffered solutions to the glass prior to the deposition of conventional powders. These supplemental interleaving materials have been found to be useful at a number of float glass facilities in the USA as well as the Orient.

A review of powder interleaving functionality was published by Düffer in 1986 [16]. This treatment provides a simple explanation of how interleaving materials serve to retard glass surface corrosion. A number of photographs and micrographs are included that illustrate how powders offer mechanical protection against scratching as well as surface corrosion.

Current Challenges Related to Glass Surface Corrosion

In spite of all the intense efforts expended by the industry in developing interleaving materials to prevent glass surface corrosion, situations develop where glass becomes susceptible to staining. Quite often, these situations arise when glass leaves the domain of the primary manufacturers in the form of finished articles produced by various glass fabricators. Typically, fabricated glass items are not protected by interleaving materials that are found on primary glass products. As a consequence, if storage of the glass is undertaken in an environment characterized by heat and humidity, surface corrosion may occur. A salient example of this scenario is shown in Figure 4. It is clear from the example shown that the challenge remains to educate fabricators regarding the phenomenon of glass surface corrosion.

Figure 4. An example of glass surface corrosion resulting from
a lack of sufficient interleaving protection.

Even with the benefit of protective interleaving in place there are several simple rules for mitigating surface corrosion in storage:

1. Prevent glass packs from getting wet by storing in doors or taking care to secure and protect packs that may be temporarily stored outdoors.
2. If glass gets wet, open affected packs as soon as possible and allow individual units to dry before restacking.
3. Warehoused glass should be selected on a first-in/first-out basis.
4. Warehouses should be well ventilated and where possible climate controlled to prevent glass from dropping to temperature below the prevailing dew point.

CONCLUSIONS

The phenomenon of glass surface corrosion has been affecting glass surface quality since the days of crown glass. However, it was not until the post-World War I era that protective interleaving systems were developed in earnest. Initially, paper materials found prominence in serving this

objective until the late 1940's when powdered materials were studied for the first time. With the advent of the float process in the 1960's, the urgency intensified for identifying a functional powdered interleaving system. In the 1970's, attention began to focus on using polymer materials as a separating material along with acidic components to control pH amidst stacked glass sheets. By 1985, supplemental application techniques were developed for working in concert with conventional powders in preventing surface staining. As late as 1995, new technologies were finding acceptance in addressing the issue of glass surface corrosion. However, in spite of all the efforts expended by primary glass manufacturers towards preventing surface corrosion, situations still arise where uniquely adverse conditions prevail that promote the occurrence of glass surface corrosion. As a consequence, there remains a distinct need to foster awareness of the phenomenon of glass surface corrosion.

REFERENCES
[1] Newton, R.G.: The Durability of Glass; Glass Technology 26 [1] 1985, 21-38

[2] US Patent 1,293,684, 2/1919; Inventor Edward B. Brogley (Mobile, Alabama) Process of Packing Glass and Preservative Composition Thereof

[3] Douglas, R. & Isard, J.O., The Action of Water and Sulfur Dioxide on Glass Surfaces, J. Soc. Glass Technol., 33 (1949) 289-335

[4] Douglas, R.W. and T.M.M. El-Shamy. Reactions of Glass with Aqueous Solutions, J. Am. Cer. Soc., **50** [1] (1967): 1-8

[5] El-Shamy, T.M.M., J. Lewins and R.W. Douglas, The Dependence on the pH of the Decomposition of Glasses by Aqueous Solution, J.Am.Cer. Soc., **13**[3] (1972): 81-87

[6] Ernsberger, F.M., Physics Chem. Glasses, 21 (1980) 146-149

[7] Düffer, Paul F. "Glass Reactivity and Corrosion," Tech Notebook for Glass and Ceramic Decorators Vol. I, Peter C. Cassabeer Editor, The Society of Glass and Ceramic Decorators, Washington DC, 1994

[8] D. E. Clark, C. G. Pantano, and L. L. Hench, "Corrosion of Glass" (Books for Industry, New York, N.Y., 1979)
[9] US Patent 2,476,145, 8/1944; Inventors J.D. Gwyn, C.B. Williams, Jr. (Libbey Owens Ford Glass Company) Protection of Surfaces

[10] US Patent 3,723,312, 3/1973: Inventor W.J. Hay, Jr. (PPG Industries, Inc.) Combined Stain Retardant and Separator Interleaving Material for Glass Sheets

[11] US Patent 4,011,359, 3/1977: Inventors G.T. Simpkin, K. Ashcroft (Pilkington Brothers Limited) Interleaving Materials Comprising Particulate Separator and Acidic Materials for Separating Glass

[12] US Patent 4,200,670, 9/1977: Inventor E.R. Albach ((Libbey Owens Ford Glass Company) Stacking Glass Sheets

[13] US Patent 4,487,807, 12/1984: Inventors P.F. Düffer, J.D. Kelly, H. Franz (PPG Industries Inc.) Mixed Acid Stain Inhibitor

[14] US Patent 4,489,106, 12/1984: Inventors P.F. Düffer, J.D. Kelly, H. Franz (PPG Industries Inc.) Two Step Interleaving Method

[15] US Patent 5,641,576, 9/1995: Inventor H. Franz (PPG Industries Inc.) Buffered Acid Interleaving for Glass Sheets

[16] Düffer, Paul F. "How to Prevent Glass Corrosion," Glass Digest **65** [12] (1986): 77-84

Environmental Issues
and New Products

MEGATRENDS IN THE COMMERCIAL GLAZING MARKET– A CHALLENGE FOR THE GLASS INDUSTRY

James J. Finley
PPG Industries, Inc.

Technology that will be essential in meeting the needs of the glazing market over the next five years and beyond will be determined by trends in the areas of energy, the environment, and aesthetics. These trends have become so universal that they have taken on the distinction of a "megatrend." Corporations study these trends to guide marketing and R&D efforts to innovate products for tomorrow's marketplace. The glass industry has been able to continuously evolve over its long history by a creative application of advances in new materials and nanotechnology to produce breakthrough technologies. These technologies have enhanced the function and value of glass and will continue to develop and provide the next generation of products. This will help the industry attain its long range goals to further reduce energy cost for both producers and consumers, provide comfort and style in building design, and bring about environmental benefits through the reduction in fossil fuel consumption and production of greenhouse gasses.

To meet the challenges of tomorrow's commercial marketplace, products employing specialty glass and coatings solutions will continue to lead the way. Glass has gone from "plain vanilla" clear to spectral selectivity or ultraclear substrates. Coatings are applied to these substrates to enhance properties and add value by providing a variety of functions including spectral selectivity, transparent conductive electrodes, and antireflectance . These products serve markets as diverse as commercial glazing and photovoltaic systems which continue to move forward at rapid pace.

This presentation gives an overview of how the glass industry is using megatrends to guide current technical efforts to meet the challenges for the tomorrow's glazing market. The focus is on products that employ coatings and specialty glass to provide solutions. These will include spectrally selective and switchable glazing and photovoltaic systems. The technology behind these products, their benefits and limitations, requirements and issues for developing these technologies, and the role the glass industry plays in their development are addressed.

The first product area of focus, spectrally selective glazing, had its origins with the introduction of performance coatings in the early 1960's. These were low transmitting and highly reflective windows with limited spectral selectivity since the transmittance was low both in the visible and infrared portions of the solar spectrum. In the early 1980's, new materials and efficient manufacturing processes were introduced to deposit thin films of metals and inorganic materials, such as oxides and nitrides, cost effectively on large area substrates. This marked a beginning of successive generations of higher performing spectrally selective glazing. Today, the drive to reduce energy consumption in buildings and meet environmental standards has led to the latest generation of spectrally selective coatings consisting of multilayer thin film stacks.

The second area is related to photovoltaics with an industry that continues to grow at an accelerated pace and where glass and coatings are playing a critically important role in the technology. The product focus for the glass industry is in the area of low iron-high transmitting glasses and transparent conductive oxide coatings to provide a durable highly transparent cover plates and electrode for solar cells. In addition, the trend towards self-sustaining or "zero" energy buildings is driving the concept of Building Integrated Photovoltaics (BIPV). This will be an opportunity for the glass industry to provide

overall integration of energy efficient (spectrally selective glazing) and energy producing (photovoltaic modules) glazing systems.

The third product area is "switchable glazing." These systems are referred to as "switchable" because they can be stimulated, e.g. by a voltage or ultraviolet light, to turn darker or lighter to accommodate lighting preferences by building occupants, and for controlling heating and cooling costs. This technology has undergone 25 years of development and while a technical success, has not met yet with widespread commercial success. There remain issues with manufacturing cost, and market penetration due to limited functionality, and lack of broad aesthetic appeal. The continued development of switchable glazing is also critically important in the overall integration of glazing systems towards the goal of zero energy buildings.

In summary, architectural glass will be driven primarily by energy, environmental, and aesthetic megatrends. Products employing specialty glass and coatings solutions will continue to lead the way. Driven by the megatrends, it is clear that the glass industry's future path will be different than any time in its history, but the opportunity for growth will also be as great.

Overview

- Megatrends – Meeting the challenge
- Three product areas that address the megatrends
- Summary and Conclusion

Copyright 2009 PPG Industries, Inc.

Megatrends

PPG IdeaScapes.
Glass · Coatings · Paint

What is a megatrend?

Definition: a large over-arching direction that shapes our lives for a decade or more.

It guides marketing and R&D efforts to innovate new products for tomorrow's marketplace.

Energy

Aesthetics

Environment

Megatrends

The three megatrends that drive the glass industry are Energy, the Environment and Aesthetics. The challenge for the industry is to produce products that serve these needs at costs that are affordable to the consumer.

Copyright 2009 PPG Industries, Inc.

Glass Industry Product Areas

PPG IdeaScapes.
Glass · Coatings · Paint

Spectrally Selective Glazing

Photovoltaics

Switchable Glazing

These three products that employ coatings and specialty glass play a critical role in addressing these challenges. The technology of these products, their current limitations, requirements and issues for further developing these technologies, and the role the glass industry plays in their development will be addressed.

Copyright 2009 PPG Industries, Inc.

Spectrally Selective Glazing

Performance coatings were introduced in the early 1960's and achieved low transmitting, highly reflective windows. In the early 1980's, thin films of metals and inorganic materials, such as metal oxides and nitrides, were applied to large area glass substrates, but were still low in transmittance. Today, the drive to reduce energy consumption in buildings and meet environmental standards has led to new generations of higher performing spectrally selective glazing. Spectrally selective coatings exhibit high LSG ratios, i.e., they admit the maximum amount of visible light (daylight) while limiting the amount of total solar energy into a building. Earlier generations of coatings lacked this property. The benefits are illustrated in the figure.

Spectrally Selective Glazing

Spectrally selective glazing, which consists of a coating on a glass substrate (line1, 2), is "ideally" designed to allow maximum visible and zero solar infrared transmittance. The spectral irradiance curve, which is the energy distribution from the sun at the earth's surface, is superimposed over the ideal spectrally selective coating to illustrate the amount of energy in each of the spectral regions.

Spectrally Selective Glazing

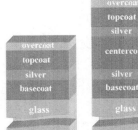  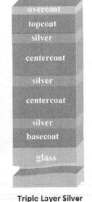

Today's spectrally selective coatings consist of alternating layers of high refractive index dielectric films (base-, center-, top-, and overcoat), and low index metal layers (silver). The double layer silver coating (center) is the most widely used spectrally selective glazing product.

The optimum coating for spectral selectivity, considering performance, aesthetics, and cost effective processing, is the triple layer silver coating (right). In addition to energy and equipment cost savings, the triple layer silver coating can also dramatically reduce the level of CO_2 emissions associated with the heating and cooling of commercial buildings.

Single Layer Silver   **Double Layer Silver**   **Triple Layer Silver**

Spectrally Selective Glazing

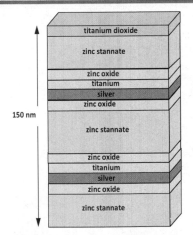 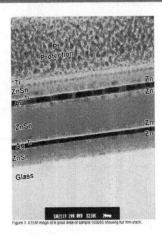

The figure on the left is a schematic of a two layer silver coating showing materials used in the layers. The total thickness of the coating is less than 150 nanometers. The photomicrograph on the right shows the two silver layers measuring about 10 nanometers surrounded by metal oxide layers.

Spectrally Selective Glazing

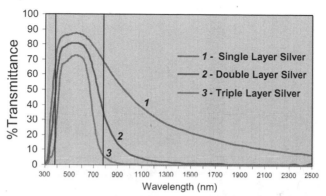

The figure shows the Transmittance for single (*1*), double (*2*) and triple (*3*) layer silver coatings on 3 mm float glass over the visible and solar infrared wavelength range. Spectrally selective coatings act to filter out the solar infrared, with more layers producing a sharper cutoff in the solar infrared and less attenuation in the visible region. This behavior gives these coatings their high LSG values, LSG being the light-to-solar gain ratio.

Spectrally Selective Glazing

Economic Impact & HVAC CO_2 Emissions Reductions

City - Boston	Annual Operating Cost Savings of vs. Tinted	Initial Capital Cost Savings vs. Tinted	Annual CO_2 Savings vs. Tinted (Tons)	40 Year Building Life CO_2 Savings vs. Tinted (Tons)
2 layer-silver	$60,474 7.1%	$203,341 8.7%	228	9,120
3 layer-silver	$97,539 11.4%	$398,881 17.1%	354	14,160

A Comparison of Energy, Economic and Environmental Benefits of Transparent Low-E Glasses, PPG Industries
Based on eight-story glass-walled office building, total Glass Area: 50,967 ft2, total Floor Area: 270,000 ft2

The table demonstrates the environmental and energy savings using spectrally selective coatings in windows versus dual-pane tinted glass for the building noted above. Double silver layer coated glass saves over 7% in annual energy costs, based on 2006 energy costs. The difference in the cost of the HVAC equipment required to heat and cool this building saves almost 9%. Triple silver layer coated glass saves almost 17% in HVAC equipment and more than 11% in annual energy costs. As far as reducing CO2 emissions, the table shows that by specifying the triple layer silver glass instead of dual pane tinted glass, 14,000 fewer tons of carbon are emitted into the atmosphere over its lifetime.

Glass Industry Product Areas

PPG *IdeaScapes.*
Glass • Coatings • Paint

Crystalline and Amorphous Si, and thin film PV are focus of Glass Industry

Market growing at accelerated rate

Glass product areas include:
TCO's
Low Iron Glass
Antireflective Coatings

Photovoltaics

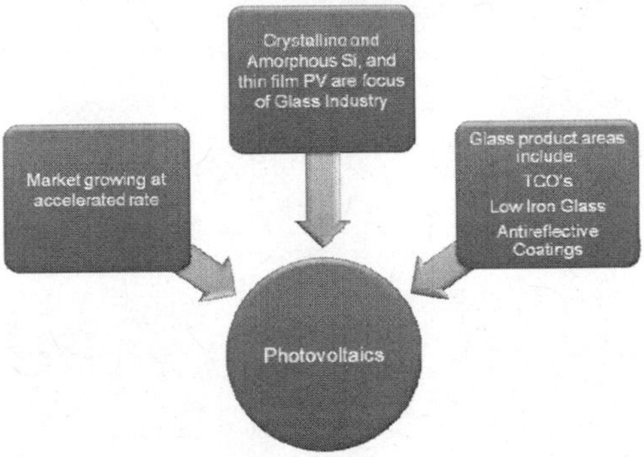

The second area of focus is photovoltaics, with glass and coatings being critically important elements in these processes. Solar power generation continues to grow at accelerated rate. Focus area for the glass industry includes transparent conductive oxide coatings (TCO's), low iron high transmission glasses, antireflective coatings, and glazing products for Building Integrated Photovoltaics (BIPV).

Copyright 2009 PPG Industries, Inc.

Worldwide PV Shipments

PPG *IdeaScapes.*
Glass • Coatings • Paint

10 Watts = 1 sq ft of glass

- Rest of World
- Europe
- Japan
- United States

Total MW Shipped

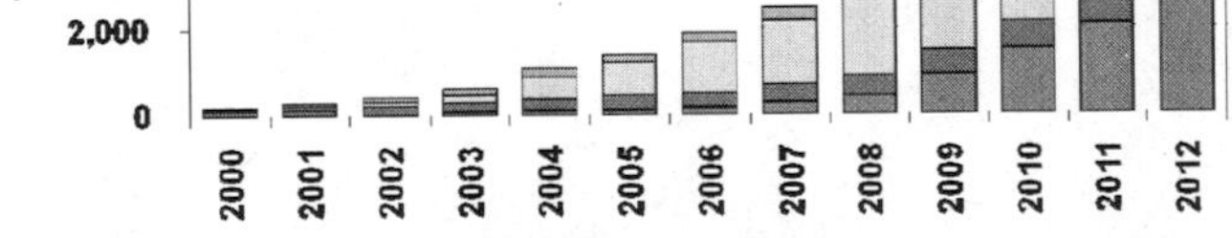

Source: Travis Bradford & Paul Maycock, Solar Power 2007, September 2007

The chart illustrates the accelerated worldwide growth of the photovoltaic market. Recently, the US has shown increasing growth after a slow start. 10 watts accounts for 1 sq ft of glass usage.

Copyright 2009 PPG Industries, Inc.

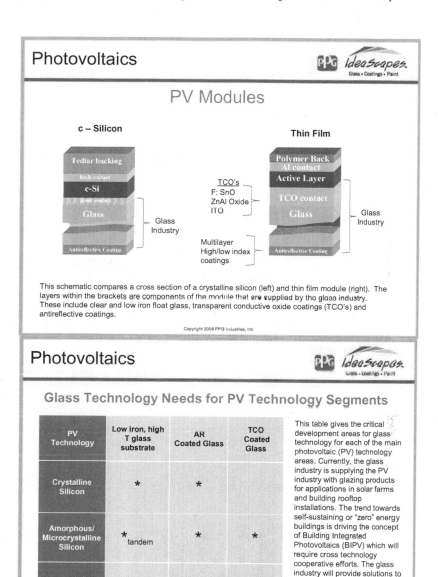

Photovoltaics

PV Modules

c – Silicon

Thin Film

TCO's
F: SnO
ZnAl Oxide
ITO

Glass
Industry

Glass
Industry

Multilayer
High/low index
coatings

This schematic compares a cross section of a crystalline silicon (left) and thin film module (right). The layers within the brackets are components of the module that are supplied by the glass industry. These include clear and low iron float glass, transparent conductive oxide coatings (TCO's) and antireflective coatings.

Copyright 2009 PPG Industries, Inc.

Photovoltaics

Glass Technology Needs for PV Technology Segments

PV Technology	Low iron, high T glass substrate	AR Coated Glass	TCO Coated Glass
Crystalline Silicon	*	*	
Amorphous/ Microcrystalline Silicon	* tandem	*	*
CdTe, CIGS Thin Film	*	*	*

This table gives the critical development areas for glass technology for each of the main photovoltaic (PV) technology areas. Currently, the glass industry is supplying the PV industry with glazing products for applications in solar farms and building rooftop installations. The trend towards self-sustaining or "zero" energy buildings is driving the concept of Building Integrated Photovoltaics (BIPV) which will require cross technology cooperative efforts. The glass industry will provide solutions to overall integration of energy efficient and energy producing glazing in the building glazing supply chain.

Copyright 2009 PPG Industries, Inc.

Glass Industry Product Areas

PPG *IdeaScapes.*
Glass • Coatings • Paint

Currently, market limited to niche applications for high shading areas, e.g., sunroofs

Respond to external stimulation: light, heat or electrical voltage

Growth depends on manufacturing efficiencies and new technology development

Switchable Glazing

The third product area is electrochromic glazing, also referred to as "switchable glazing." While a technical success after 25 years in development, it has not had broad commercial success. There remain issues with manufacturing cost, and market penetration due to limited functionality, and lack of broad aesthetic appeal. These systems are referred to as "switchable glazing" because they can be stimulated, e.g. by a voltage or ultraviolet light, to turn darker or lighter to accommodate lighting preferences by building occupants, and for controlling heating and cooling costs.

Copyright 2009 PPG Industries, Inc.

Switchable Glazing

PPG *IdeaScapes.*
Glass • Coatings • Paint

1 - Clear State
2 - Switched State – High T
3 - Switched State – Low T

The chart illustrates "ideal" behavior of a switchable device for three different states in the visible and infrared regions of the solar spectrum. The clear state (1) allows transmittance across the solar spectrum. States 2 (High T) and 3 (Low T) limit the transmittance in the infrared and allow high (2) and low (3) transmittance in the visible region of the spectrum. Ideally, the visible spectrum can be varied from 0 to the maximum glass transmittance, and the infrared can be varied independently of the visible spectrum to optimize energy savings. There are no devices in existence today that have this capability. The spectral irradiance curve, which is the energy distribution from the sun at the earth's surface, is superimposed over the ideal curves for the device to show the relative amount of energy in each of the spectral regions.

EC Device & Circuit

Electrochromic Device Features

Device

* Transparent conductive oxide (TCO) coated glass acts as an electrode (supplied by the glass industry)
* Coloring anode and cathode (polymer, dye, organic) provides visible and solar IR shading
* Electrolyte (liquid or solid state) supplies electrical charge to darken and lighten anode and cathode
* Voltage cycles electrical charge to cathode and anode
* Power Supply/Control Unit provides voltage to power and switch device
* Sensors (not shown) adjust device to environmental conditions

Copyright 2009 PPG Industries, Inc.

Electrochromatic Glazing

Tungsten Oxide (WO₃) Based Cell

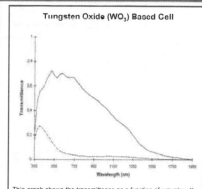

This graph shows the transmittance as a function of wavelength in the clear and darkened state for an electrochromic device using thin film WO3 as the coloring electrode The chart illustrates the simultaneous change in transmittance in the visible and solar infrared portions of the spectrum from the lightened to darkened state. WO3 based devices have been successfully commercialized for large area EC glazing such as skylights.

Other Similar Switchable Technologies

* Liquid crystals
  – not likely to be used for building applications

* Thermochromic & Photochromic
  – currently not viable or cost effective in commercial applications

* Suspended particle display (SPD).
  – scatters light in the off-state and clear in the energized state
  – seen limited commercialization after several decades

* Combined electrochromic and photovoltaic windows
  – explored in R&D labs
  – only at the concept stage

Copyright 2009 PPG Industries, Inc.

Summary & Conclusion

Summary

• Architectural glass will be driven primarily by energy, environmental, and aesthetic megatrends.
  – Products combining specialty glass and nanotechnology solutions will continue to lead the way.
  – Energy efficiency will be guided by integrated design processes (e.g., BIM) and measurement and feedback of real world systems.
• The industry focus will be on
  – added value products that optimize energy efficiency and provide aesthetics,
  – improved manufacturing efficiencies.

Conclusion

• The opportunity for growth will be great and driven primarily by energy, environmental, and aesthetic megatrends.
• The glass industry's future path will see more change over the next five years than at any time in its long history.

LOOKING WINDWARD: FIBER GLASS IN THE ENERGY MARKETS

Cheryl A. Richards
Global Market Development Manager, Wind Energy
PPG Industries

As a marketer, I appreciate that when in a difficult situation we call something unique or an opportunity, so shall we call our Glass Problems by a new name, glass opportunities instead? I want to take show you a few markets that are providing growth in fiber glass

As you know through your involvement in the US Glass industry, Glass Problems Conference, and GMIC, the glass industry is searching for energy efficiency in our melting processes to manage energy cost and reduce emissions to lighten the environmental footprint. This discussion focuses on one glass industry product, fiber glass, and how it is helping with energy production. I will start our discussion a brief overview of the fiber glass market and step through a few energy applications. I will conclude with an overview of why fiber glass works in these applications the opportunity for new products.

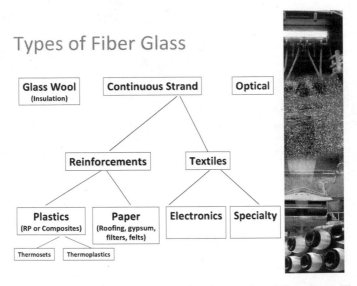

The U.S. Department of Commerce identifies three broad categories of fiber glass. The largest segment in terms of volume, dollar value and familiarity to the public is the glass wool insulation business. There are far more fiber glass insulation plants in the U.S. than any other type because the product is so light and voluminous and regional production is the most viable way to compete. Industry leaders are Owens Corning, Johns-Manville and CertainTeed (Saint Gobain). Continuous

strand fiber glass is the portion of the industry in which PPG participates and you can see how this category breaks down into reinforcements for plastics (or composites) and reinforcements for other materials like wet chop which is converted into a paper-like reinforcement for asphalt roofing shingles or vinyl flooring materials. The RP or composites business is commonly broken down by resin family, either Thermoset or Thermoplastic. Fiber glass textiles or yarn are commonly converted into fabrics used in the electronics industry or for a wide variety of other specialty applications. Industry leaders are OCV, PPG, JM and several up and coming Chinese producers. The third and final category of fiber glass is optical fiber glass. It has a different chemistry and a very different manufacturing process. One industry leader in this category is Corning.

Fiber Glass Markets
GLOBAL BREAKDOWN BY END USE

Aviation, Military & Ballistics 1.5%
Electrical & Business Equipment 6.4%
Other 1.9%
Consumer & Appliance 7.3%
Infrastructure 8.8%
Transportation 27%
Corrosion & Industrial 14.9%
Marine 4.7%
Construction 27.7%

Global fiber glass demand for composites was 7.5 billion lbs in 2007 with net sales of US$4.6 billion

Source: PPG Industries

This chart shows PPG's estimate of fiber glass demand according to end use segments or where the product ends up. From a marketing perspective, we need to be concerned with market size and growth rates. I will highlight the growth rates in key segments and applications as we move forward. First some of the markets for composites include:

- Construction – lineals, doors, shingle, bridge
- Transportation – air intake manifold, underhood, interiors dashboards, hoods, battery trays, shroud, housings, truck hoods and cabs
- Aviation/military – interiors, overhead compartment, up armoring of vehicles and body
- Electrical – FR4 circuit boards, cable trays
- Consumer Appliances – skis, personal craft, handles, displays, housings

As for growth, reinforcements have experienced just under a 10% growth rate from 2004-2008 which compares favorably to a world growth domestic product of 3.5% for the same time frame. The

projected growth rate for composites for the next five years is just under 9% versus a projection of 2% for world GDP.

Our focus for this discussion will be the segments called infrastructure and corrosion & industrial. The infrastructure segment includes end uses such as wind energy, gratings, and bridges. Corrosion & industrial contains infrastructure for oil, gas, compressed natural gas (CNG), coal as well as tanks, piping, flues, ducts, and cooling towers. Let's start with wind energy.

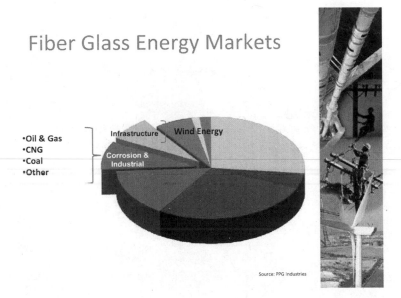

Fiber Glass Energy Markets

•Oil & Gas
•CNG
•Coal
•Other

Infrastructure   Wind Energy

Corrosion & Industrial

Source: PPG Industries

Most people are familiar with wind energy. Let's start by defining some terminology. A wind turbine is typically composed of three blades, a generator contained in a nacelle, and a tower. Typically we describe this market by how many megawatts (MW) or towers are installed. That information allows one to calculate market size. There are a few sizes going into the market which includes: Small – homes, farms, and remote applications (e.g. water pumping, telecom sites, ice making); Intermediate - village power, hybrid systems, distributed power; and Large - central station wind farms, distributed power.

Wind Energy

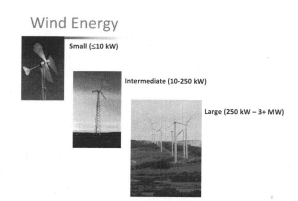

Small (≤10 kW)

Intermediate (10-250 kW)

Large (250 kW – 3+ MW)

The market which is getting the most attention is the large category or those dedicated to a wind farms. The blades which capture the wind and allow the turbine to convert the wind to electrical energy are made of fiber glass reinforced plastics. This is one of the largest composite structures in the market.

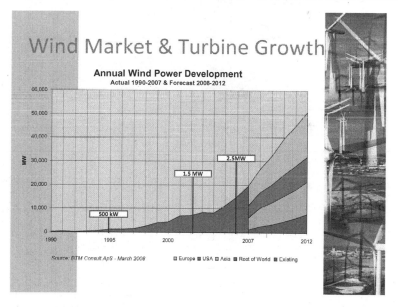

This graph represents the historic and forecasted growth rate. The market has a compounded annual growth rate of 25% for the past five years. As you can see, the graph shows capacity going in all major global regions. This means that more manufactured components and fiber glass production is

required in those regions to meet this high market growth rate. The projected growth rate shows that this is a desirable market for fiber glass well into the future. The second thing that is illustrated is that over the past two decades, the size of these units has increased. Starting from blades that were less than 10 meters in length to 39 meter blades used in the 1.5 MW units and now the 61.5 meter blade for the 5 MW units. In this segment we want products with higher mechanical performance to enable the larger blade size and to extend service life.

Wind Blades
Increasing Power

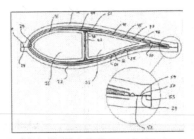

Source: Siemens, RePower

Note that the photo of a cross-section of a blade, which is composed of a skin, spar, spar caps, demonstrates that this is a complex structure. Wind blades are designed to several major structural conditions. These include tip deflection, strength, and buckling during severe loading and blades must endure a high number of fatigue cycles which vary between tension and compression. What is important are the requirements of the fiber glass in the composite. To the blade designer, fiber glass strength and stiffness are critical as well as how the fiber glass performs in the composite, or properties such as fatigue and durability, where the fiber glass has been optimized for resin of choice.

Composites are also used directly and indirectly in the extraction and distribution of oil and natural gas. Shown in this series of photographs are applications for tubing and plastic piping for oil and chemical transport and composite tanks for storage of raw oil. We have come a long way when you consider that the first pipe was made out of a hollowed log. As a matter of fact, the first application for gas and oil pipe was to transport gas to lanterns in London beginning in 1806. The first oil well in the US, drilled in Titusville, PA in 1859 by Col. Edwin Drake, who transported the oil in old wine barrels to market. What you may not know about oil and gas infrastructure is that mature fields produce a higher water cut when the easy to extract oil is depleted. Produced water has a high chlorine content and is corrosive to steel piping. Corrosion is an area where composites excel. When enhanced recovery methods are employed the well is injected with produced water back into formation to flush oil out. Tertiary production follows and the use of CO_2 is employed. CO_2 is miscible with oil and blended with the oil to make it flow. Wet CO_2 is very corrosive.

Oil and Gas Infrastructure

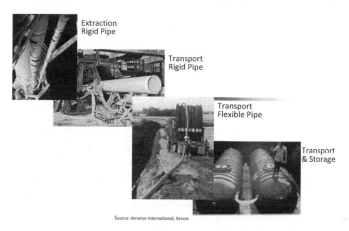

Source: Ameron International, Xerxes

Interest in this the oil and gas infrastructure market comes back to demand. Recently we have experienced an increase in oil and gas demand driven by world markets. As a result, there is an increase in demand for composites which is driven by increased need for corrosion resistance as I previously explained and decreased availability of traditional materials (steel) in the piping industry. Composite growth in this market is further driven by the increase in price for carbon steel which brings the price for both steel and composite pipe close. Interestingly, steel pricing has been much more volatile and has seen higher increases than fiber glass composite pricing. A projected growth rate for composite applications in this market at better than 10% per year. An interesting fact is that total mileage of pipeline in 2000 was about 160,000 miles of oil pipeline, 300,000 miles of natural gas main lines and 1.9 million miles of natural gas distribution pipelines. Composites makes up less that 2% of this mileage. So, as oil fields mature and extraction methods get more sophisticated and more water is produced in combination with oil corrosion precludes the use of bare carbon steel.

Oil and Gas - Drilling into it

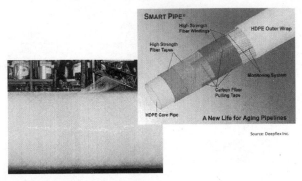

Source: Deepflex Inc.

In oil and gas drilling, the first photograph shows the filament winding process which is a considered an 'industrialized' composites process used to make pipe and tanks. This photo shows a double helical process which maximizes the longitudinal and circumferential strength. In the photograph on the right, the SMART PIPE® is made by an extrusion type process. There is a high use rate of fiber glass is these applications of about 70% glass in either polyester, vinyl ester or epoxy resin systems. Pipe requirements include life of around 20 years, ability to sustain temperatures of up to 150°F, ability to withstand a corrosive environment (salt water), ability to take pressures from gravity fed up to 3500 psi, and exposure to fatigue or cyclic pressure variation.

Advantages of fiber glass include high corrosion resistance, lightweight, high tensile strength and modulus and as a composite high fatigue life. Other considerations for composite pipe include installation locations (lightweight), easy of repair (sectional), and wear resistance. In case of the SMART® or flex pipe, there are no joints and this pipe is spoolable and can be installed at a rate of 10000ft/day.

Compressed Natural Gas

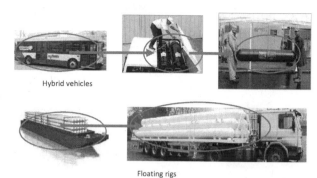

Hybrid vehicles

Floating rigs

Source: Lincoln Composites, Floating Pipeline Co.

The compressed natural gas (CNG) market consists of high pressure gas storage and transport for alternative fueled vehicles, both natural gas and hydrogen. This includes OEM vehicles and buses. Most vehicles use steel tanks, but there is a movement towards composites for weight reduction in buses. The US market uses CNG vehicles in public transportation. In city buses, 25% of orders are for CNG fueled buses. Use of carbon fiber is necessary for roof mounted vehicles. There are two to four tanks per pack in buses. In the case of gas transport, marine or floating rigs are real! In both applications corrosion resistance is critical, but other properties which drive fiber glass or composites use in these applications are light weight, ability to withstand high internal pressure and high impact resistance.

In the case of wind energy fiber glass demand is driven by turbine builds. In the case of oil and gas, we can look consider fuel consumption and assume a percent of the supporting infrastructure employs the use of composites. For CNG, we can use a similar approach but it is not as clear. Natural gas is the cleanest and most widely used and available alternative transportation fuel. There are greater than five million CNG vehicles world-wide. Regions showing high growth for OEM vehicles and buses include Europe, South America, and Asia. A commitment by government is required for infrastructure for fueling to foster market growth. As most vehicles use steel tanks, a move towards composites, which afford weight reduction, is occurring.

There are four basic types of CNG tank designs. The design used depends on need to reduce weight and how much the user can pay. All designs have equivalent safety, as all must meet the same standards. Type 1 is all metal which is cheap but heavy. Type 2 has a metal liner reinforced by composite wrap, either glass or carbon fiber, around middle. This type is less heavy but more costly. Type 3 has a metal liner reinforced by composite wrapped around the entire tank which is light-weight, but expensive. Type 4 tanks are composed of a plastic gas-tight liner reinforced by a composite wrap around the entire tank. So composites made from fiber glass in this market offer a light-weight structure, high corrosion resistance, and safety from rupture. Composite tanks are less costly than

comparable steel pressure vessels. It is worth noting that the global price of steel has more than doubled over the past several years, making composites even more cost competitive.

Coal Infrastructure

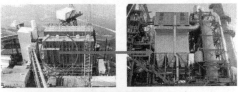

Baghouse

Stack liner and scrubber

Source: Alstom, Ershigs

In the case of coal fired generation, the last area of focus, fiber glass is used in infrastructure. The top photo shows a bag house and the bottom applications are stack liners and scrubbers. The bag house uses different fiber glass products than those we have been discussing (composites). Fiber glass in these applications is knitted or woven, are not composites and are known as filters. In the US, coal plants emit 48 tons of mercury per year. Coal is a desirable form of energy generation since it is low cost and has long term availability. But regulation of emissions is driving the higher requirement of filter particulate matter from the effluent.

It is difficult to say what the growth rate in this market is but the compounded annual growth rate for coal fired power generation is between 3 -13%. A reasonable estimate is at the lower end or around 6% based on cancellations due to environmental reasons. As mentioned previously, the low cost of coal, long term availability, and expansion of existing power plants is driving this trend. However the market size is impacted by something not seen in this curve, and that is existing power plants must be retro fitted with equipment to comply with tighter environmental regulations.

Coal – Fiber Glass Can Take the Heat

Source: Ershigs

The top picture is a close up of a bag used in a bag house. There can be up to 10,000 tubes per bag house. Its purpose is to remove particulate matter. The bottom picture shows the lining of a stack which is the traditional fiber glass composites that we have discussed. Requirements for the fiber glass in bag house applications are ability to withstand operating temperatures of up to 350°F, ability to function for 36 months to 10 years and ability to withstand acid corrosion (H_2SO_4). The stack liner is protecting the stack asset which can be composed of cement or steel. In the case of the stack liners, the temperature requirements are less than that for the bag or 200°F. Other properties discussed earlier, corrosion resistance and light weight, are what drives fiber glass use in stacks.

E Glass – Current Attributes

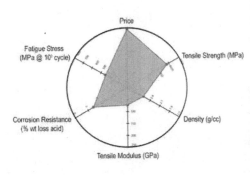

Ninety to 95% of fiber glass sold is E fiber glass, which is defined by a specific chemistry range. The above spider graph shows a map of the E fiber glass space. The important properties of fiber glass or composites are shown on five of the axis. These are not all of the important properties that are inherent in E glass, but are key properties to the markets reviewed in this presentation. Also shown is price. E fiber glass has been around for more than fifty years and has been optimized for low cost production and high mechanical properties within the composition range.

Carbon Fiber – Fiber Glass Comparison

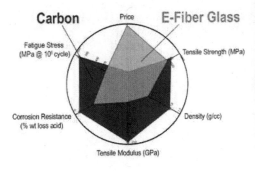

The above spider graph is a comparison of E fiber glass to carbon fiber. Carbon fiber in terms of its low density (light weight) and high tensile modulus is a viable competitive fiber. As wind blades get larger and need to maintain their light-weight attribute or in high pressure CNG tanks carbon fiber can

be a material of choice. The disadvantage with carbon fiber includes its high price and low availability, which is not plotted.

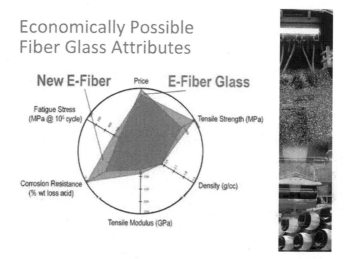

Based on these new markets for fiber glass composites, which have high growth rates and demand high volumes world-wide, we can consider the possibility to optimize the fiber glass for improved properties. That is properties that are closer to that of carbon fiber but still offer a competitive price in the market place and produced by a process that is easily industrialized. Taking fiber glass to the next level is not a problem, but an opportunity, to be developed by those in research, development and engineering. It is an opportunity that requires new batch materials, innovative glass chemistries, efficient melting processes and new melting technologies, as well as the engineers and scientists to nurture that future state.

CURRENT GLASS FURNACE AIR EMISSION COMPLIANCE ISSUES

C. Philip Ross
Creative Opportunities, Inc.
Laguna Niguel, California

INTRODUCTION

Over the past three decades, air emission regulations have influenced the glass industry's decisions regarding the design and operation of their melting furnaces to comply with increasingly stringent requirements. This presentation will provide an overview of the current and proposed regulations in the U.S., and attempt to address some of the issues facing glass furnace operators.

The cost of compliance has become a significant issue for all glass manufacturers, and new environmental regulations continues to burden glass manufacturers financially, requiring the purchase of add-on devices, making process modifications to existing units, or total unit replacement. To comply, manufacturers encounter significant costs for capital investment (often with no increase in production), operating cost penalties (energy, production or inefficiencies), monitoring equipment, labor and other maintenance costs. Cost of compliance varies, depending on method of control selected, the level of reduction to be obtained, and the way in which measures are integrated into the operation of the furnace.

GLASS FURNACE AIR EMISSIONS

Environmental compliance for criteria air pollutants from glass melting include emissions of fine particulate (< 10 microns), NO_x, SO_x, CO, VOC's, heavy metals, and (in the near future) greenhouse gases. Each of these emission species are generally being regulated by state, regional, or local agencies, typically mandated by amendments of the Federal Clean Air Act (CAAA). The USEPA has determined glass plants emit approximately 174,000 Tons per Year of NO_x, SO_2 and Particulate Matter, and most furnaces are considered "major" emission sources.

Air emissions from glass melting furnaces are a result of both the high temperature combustion process and some of the glass chemical components evolving out of the melt process.

Products of Combustion Gases

NO_X - An inherent by product where Nitrogen is present within the combustion process. NO_X is actually NO (but considered as NO_2 for emission rate calculations)

CO - A remnant of incomplete combustion (often associated with NO_X reduction techniques)

SO_X - Gas byproduct from sulfur containing fuels (oils, per coke)

Batch Raw Material Ingredient Evolution

SO_X - The evolution product from the refining process of sulfur containing raw materials (Sulfates, Sulfides, & Sulfur in Cullet)

Particulate - Condensable solid compounds, created within the cooling exhaust gases, evolved as gases from the glass melt (including Alkali, Borates, Sulfates, and Carbonates)

Heavy Metals - Compounds from low vapor pressure metals (such as Arsenic, Lead, Chrome, etc.)

ENVIRONMENTAL REGULATORY BACKGROUND

The first significant emission standards for U.S. glass furnaces began in the 1960's, with limits of stack opacity (measured visually). Test methods were subsequently developed to measure the weight of this solid particulate, and regional or state agencies (such as Los Angeles and New Jersey) began to set mass limits (pounds/hour). In the 1970's many furnaces were fired with fuel oil, and some regions placed limits on SO_X emissions or limited the amount of sulfur in the fuel.

Until 1979, there were no Federal (National) air emission standards for glass furnaces. Then, the U.S. EPA developed a New Source Performance Standards (**NSPS**) standard for <u>particulate</u> emissions from all new or "modified" glass melting furnaces.[1] The standard for a new furnace (0.2 pounds per ton of glass melted for container glass) effectively required the use of a particulate capture "add-on" device (i.e. "ESP" Electrostatic Precipitator, fabric filter Bag House, or a wet Scrubber). The standard was slightly more lenient for oil fired furnaces, and certain glass types - such as borosilicates. This regulation incorporated a precedent for the standard to be based upon mass (gm per kg or lbs. per ton) vs. concentration (mg/nM^3) based limits similar to Europe.

The final rule established conditions which furnace repairs could trigger applicability of "new source" requirements, including a maintenance cost vs. replacement cost test. Existing furnaces which were "modified" or "reconstructed" (typically during a rebuild where the cost exceeded 50% of the replacement value) would be required to also meet a less stringent standard (1.0 pounds per ton for container glass). This standard might still require add-on technology, or process modifications such as high levels of electric boosting or greater recycled cullet content.

The container industry was successful in convincing EPA to specifically exclude "rebricking" and the cost of rebricking from the definition of "reconstruction" under the Part 60 regulations. This provision protected existing furnaces from triggering more stringent ("modified") emission standards at the time of cold repair (rebricking). Consequently most container furnaces in the U.S. still do not have add-on particulate control equipment. Until recently, this was because most existing container furnaces have avoided being considered "modified" - unless their size and production capacity was increased.

Concerns with Ozone / Smog triggered standards for NO_X and VOC's as precursors - initially in California. In the early 1980's, the California Air Resources Board (CARB) established a NO_X Model Rule. In response the Los Angeles, San Francisco Bay, and San Joaquin Valley areas set NO_X limits for glass furnaces. Prior to 1990, only these regions in California had specific NO_X regulations.

The Federal Clean Air Act Amendment of 1990 (CAAA) set ambient air quality standards for the entire country. Based upon the level of emissions, stationary sources above typically 100 tons per year were deemed "Major" sources and subject to new rule making at the State or Regional level. There are approximately 20,000 sources that would be classified as major under the Clean Air Act, and most of the 375 glass melting furnaces in the U.S. are considered "major" sources.

Operating permits under Title V of the act established emission limits, along with "compliance assurance" monitoring record keeping, which became Federally enforceable. Industry specific rules

---

[1] (40 CFR Part 60, Subpart CC)

and limits were subsequently established, with the regions with the worst air quality requiring the most severe reductions in emission standards. This era significantly contributed to the adoption of oxy-fuel firing to address NO_X emissions from glass furnaces. In some regions, such as Los Angeles, emission reduction credits (quantified with CEMS) could be sold and offset the cost of "over controlling" emissions through the "RECLAIM" trading program..

In 1996 the U.S. Environmental Protection Agency proposed more stringent national air quality standards for particulate matter (including the new category, PM $_{2.5}$) and ground level Ozone (smog). More regions became considered out of compliance with the Ambient Air Quality Standards. This has recently triggered the National Emission Standards for Hazardous Air Pollutant (NESHAP) requirement for meeting the NSPS 0.2 lb. per ton particulate standard for wine glass producers (Chromic Oxide trigger), and other glasses using any of seven heavy metals. Since 2000, regions such as the North East and Midwest are being mandated to reduce emissions which travel between states. This presently includes NO_X and SO_X. Concerns with "urban haze" is anticipated to force even tighter requirements on all PM $_{2.5}$ particulate precursors (NO_X, SO_X, VOC's), such as this year's actions in the San Joaquin Valley in California.

CURRENT STATUS OF EMISSION REGULATIONS

State and Regional authorities address the level of not attainment for their ambient air quality and determine the amount of emission reductions required from all sources. Typically, the existing total emission inventory (collected from operating permit limits) is compared to some reduced level, which will be necessary for the ambient air quality to come into compliance. The areas with the greatest need to reduce emissions will require the most demanding emission reductions. Regulations are then developed for each source category, which are based upon technical and economic factors. Each regulatory authority can interpret what is considered the Best Available Control Technology (BACT), but in practice most are adopting definitions from other regions (typically California).

Particulate matter (PM) is a mixture of small solid particles and liquid droplets found in the air. EPA's original PM ambient standard called for regulation of particles the less than 10 microns (PM_{10}) on size or in concentrations of 50 micrograms per cubic meter annually and 150 micrograms per cubic meter daily. The new PM standard calls for 2.5 microns or smaller (PM $_{2.5}$) in concentrations of 15 micrograms per cubic meter annually and only 50 micrograms per cubic meter daily. Essentially, the size of all particulate from glass furnaces is less than 2.5 microns.

EPA considers that $PM_{2.5}$ can also be formed in the atmosphere from "precursor" gases such as sulfur dioxide, nitrogen oxides, and volatile organic compounds. Presently furnaces are judged by physical total solid particulate, < 10 microns, and "condensable" particulate. A specific source test method for PM $_{2.5}$ has not yet been approved by EPA.

SO_X - Generally there are have been no SO_X emission limits for glass furnaces in the U.S. This is because they are now mostly fired with natural gas. The primary source of SO_X emissions in glass furnaces come from raw material Sulfates in the batch formulation, which are also a major contributor of Particulate in Soda-Lime glasses. Present practice for Soda-Lime is to use low sulfate batches, and consequently SO_X emissions are less than 2.0 lb. / ton. The San Joaquin Valley is currently establishing a rule modification for a SO_X emission limit of 0.9 pounds per ton of glass for container furnaces.

NO_X - Ozone is the prime ingredient of urban smog, and ambient air quality standards are being regulated. Ozone is not emitted directly into the air, but rather is formed by reacting gases of nitrogen oxides (NO_X) and volatile organic compounds (VOC's), which in the presence of heat and sunlight, react to form atmospheric ozone, particularly during hot weather. NO_X is emitted from motor vehicles, power plants and most other sources involving combustion. VOC's are emitted from a variety of sources, including motor vehicles, chemical plants, refineries, factories, consumer and commercial products, and other industrial sources.

The Federal Clean Air Act Amendment of 1990 established ambient air quality standards for Ozone. Regions not in compliance with the standard are required to regulate NO_X and VOC emitters (including glass furnaces). Areas with worse quality air are setting more stringent standards. Container and Flat Glass segments have furnaces in a number of effected regions, and the following limits are specifically in effect.

Container	lb. / Ton
New Jersey, Pennsylvania, San Francisco	4.0
San Joaquin Valley, Ca.	4.0 (then 1.3 after 2011)
Los Angeles, Ca.	< 1.5 annual reductions to 0.86 by 2010

CO / VOC - In the past, glass furnaces had essentially no CO limits The latest standard in California's San Joaquin Valley imposes a 300 ppmv limit for CO and 20 ppmv for VOC's in their Rule 4354.

GLASS INDUSTRY PREFERRED EMISSION CONTROL TECHNOLOGIES

The methods actually employed for reaching compliance with environmental regulations will depend upon many factors - including specific furnace design, site constraints for available space, relative capital investment vs. operating costs, health risks relating to labor and maintenance requirements, and the potential impact upon glass quality or production efficiencies. Some of the "process" control technologies (not involved with add-on equipment in the exhaust stream) being employed on glass furnaces to obtain modest emission reductions include -

NO_X - Excess O_2 control, low velocity burners, fuel or air staging combustion (OEAS)
SO_X - Batch Sulfate alternatives, avoiding Sulfur containing Fuels
Particulate - Higher Cullet levels, lower batch Sulfates, lower temperatures with electric boosting

Add-ons for particulate control include fabric filter Bag Houses, Electrostatic Precipitators (ESP), or wet Scrubbers. Wet scrubbers have high operating costs and liquid waste disposals concerns, and have essentially been ruled out by the glass industry. Bag Houses have lower capital costs than ESP, but higher operating costs. Bag Houses have slightly better collection efficiency than ESP's, but maintenance and reliability generally offset this advantage.

SO_X add on control involves injecting a "wet" or "dry" chemical reagent into the hot exhaust stream to combine with gaseous SO_X and form additional particulate, which is then typically captured in a downstream ESP. Typical reagents used include powered $Ca(OH)_2$, a water solution with Trona, Soda Ash or Caustic Soda. The captured sulfate particulate can be potentially recycled in the batch, preferably after the fine particles have been agglomerated back to the size of the other batch ingredients.

The ultimate NO_X reducing technology for glass melting is an all-electric furnace. However, this is not typically used because of the inherently shorter refractory life, color and cullet chemistry restrictions, a narrow operating pull rate range ("turn down ratio"). Oxy-fuel firing is presently the U.S. glass producer's first choice for meeting low NO_X emission standards. European practice continues to stay with more traditional air combustion furnaces - which incorporate Selective Catalytic Reduction (SCR) after a SO_X scrubber and ESP. One U.S. float furnace is currently operating in California with this technology. It is an alternative to Oxy-gas used by their competitor for NO_X control, in the same District. Moderate (> 50 %) NO_X reduction has also being obtained with Pilkington's 3R process or Ammonia injection without catalyst Selective Non Catalytic Reduction (SNCR) in four California furnaces. Air Staging (OEAS, BAAS) is a retrofit technology adopted by a number of manufacturers.

PERMITTING / RE-PERMITTING ISSUES

Since enactment of the Federal NSPS in 1979, there have only been five new (green field) glass container furnaces in the U.S., and all these have required particulate capture equipment. Presently there are a total of 117 container furnaces in the U.S., and only ~10 percent of the furnaces (all built before 1979) now have particulate control add-on equipment. As all of these furnaces have undergone cold repair outages and most have increased their pull rates, EPA is exerting significant pressure to determine if these furnaces should be meeting the more stringent NSPS requirements as "modified" sources. This issue involves a number of definitions and interpretations.

NSR / PSD Overview

The basic requirements of the *"New Source Review" (NSR)* program are established in parts C and D of Title I of the Clean Air Act. NSR typically refers to the construction permit program for "major" sources. NSR's purpose is to ensure that 1) air quality does not worsen where the air is currently unhealthy to breathe, and 2) air quality is not significantly degraded where the air is currently clean. The fundamental philosophy underlying NSR is that a source should install modern pollution control equipment - when it is built (for new sources) or when it makes a major modification (for existing sources) that increases emissions significantly.

The major NSR program is actually comprised of two separate programs: 1) "Non Attainment NSR" and 2) "Prevention of Significant Deterioration (PSD)". These two programs have separate requirements for addressing the varying air quality plan needs in different areas where they apply.

Non attainment NSR applies in areas where the established NAAQS for a Clean Air Act pollutant is not being met. These areas are called "non attainment regions". Non attainment NSR for major sources of certain pollutants also applies in the federally designated ozone transport region (OTR), which consists of eleven Northeastern states[2].

PSD applies to major sources located in areas where air quality is currently acceptable – i.e. where the NAAQS for a Clean Air Act pollutant is being met. These are called "attainment regions". Because non attainment areas have poorer air quality, non attainment NSR requirements are generally more stringent than PSD requirements.

---

[2]Connecticut, Delaware, Maine, Maryland, Massachusetts, New Hampshire, New Jersey, New York, Pennsylvania, Rhode Island, Vermont, and Washington, D.C.

EPA believes that all existing emission sources are aging and will have to be replaced, and the replacements will automatically have to meet the new source emission standards when repaired. By EPA's count ~8-900 existing major sources should be triggering PSD/NSR permits annually. Since the actual number is only ~200, EPA is trying to change the past interpretation of rules. They are now taking the position that, because of their interpretation of PSD or NSR requirements for permit applications, all major sources should be required to eventually meet NSPS emission standards.

NSR Applicability Triggers & Exemptions

NSR regulations require "major modifications" be permitted under the PSD and NSR programs. However, NSR only applies if the construction project results in the *potential to emit* air pollution in excess of certain threshold levels established in the NSR regulations. For a *new source*, NSR is triggered only if the potential emissions qualify as "major." For an *existing major source* making a modification, NSR is triggered only if the *modification* will result in a significant increase in emissions.

There are certain activities that are exempt from NSR. They are defined in the regulations as *exclusions* from the definition of a physical change or change in the method of operation. For example, a specifically defined *routine change* is exempt from NSR. In 1978, the Glass Packaging Institute had sent a position paper to EPA with the definition of "**reconstruction**" for glass container furnaces as "routine". This position paper was part of GPI's effort to ensure that the cost of "rebricking" was not included in the definition of "reconstruction" under the Glass Manufacturing NSPS. It is significant that their language regarding routine practices was incorporated into the final NSPS document.

To appreciate the current issues faced by glass manufacturers when making furnace changes, there are specific terms and interpretations which require review. The term "**major modification**" is generally defined as a "physical change or change in method of operation" of a major source that "would result in a significant net emissions increase" of any regulated pollutant. Factors to consider include:

1. Whether a particular project falls within the exclusion from the definition of "physical change" for activities that constitute "**routine maintenance, repair and replacement**" ?

2. Whether a physical change resulted in a "significant net emissions increase"? In particular, whether EPA may apply an "**actual-to-potential**" emissions methodology, or whether EPA is required to make an "**actual-to-actual**" emissions comparison.

3. Whether EPA has the authority to **aggregate separate construction projects** into a single project in determining NSR applicability?

4. Whether emissions changes at "**debottlenecked**" units can be considered in determining NSR applicability and, if so, what emissions comparison methodology (actual-to-actual, potential-to-potential, or actual-to-potential) is to be used?

Because the lack of growth in the container glass industry, very few new glass furnaces have been constructed in the U.S. since 1980[3]. Rebricking and / or modifications have been accomplished, and emission limits have been established to meet the Federal NSPS. EPA now wants to challenge

---

[3]In 1981 there were 129 glass container facilities in the U.S. Presently there are 54.

industry's practices of *"routine maintenance, repair and replacement"* by look at other processing equipment associated with upgrading production capabilities. On December 31, 2002 the Environmental Protection Agency (EPA) published its final revisions to the New Source Review (NSR) programs mandated for both attainment and NonAttainment areas[4]. These revisions include the subject matter of *Baseline Emissions* determinations and *Actual to Future-Actual* Methodology.

"Major Modifications" - A modification is defined in 40 CFR § 60.14(a) as "any physical or operational change to an existing facility which results in an increase in the emission rate to the atmosphere of any pollutant to which a standard applies." For NSPS purposes, an increase in emission rate is determined in terms of kilograms **per hour**. The change in emission rate associated with a physical or operational change is determined by comparing the hourly emissions at maximum capacity prior to the change with the hourly emissions at maximum capacity after the change.

"Routine Maintenance, Repair and Replacement" - Under EPA's rules, these activities are excluded from the meaning of "physical change" and thus are exempted from NSR applicability. The glass industry has statements by EPA on the record regarding what constitutes routine maintenance. EPA has confirmed in the past that rebricking is routine for the glass industry. For example, in the 1980 Final NSPS Preamble, EPA explained that "[t]he rebricking exemption was not questioned due to the regularity and necessity of the operation to this industry." Recent EPA enforcement statements make clear that EPA believes that the scope of the exclusion should be limited to "frequent, traditional and comparatively inexpensive" activities. EPA often applies the traditional "four-factor" test, looking at 1) the nature and extent, 2) purpose, 3) frequency, and 4) cost of the project.

Physical Changes - The methodology for determining whether a physical change results in a "significant net emissions increase." The term "net emissions increase" is defined in the regulations as "any increase in actual emissions" from the change. Some interpretations by EPA of the routine repair and replacement exemption would only cover a "brick for brick" replacement. Any other design or construction change might trigger the requirement to look at a furnace rebuild status. Although a physical change may have occurred to the melter, the changes not be considered modifications with respect to NSPS subpart CC if the operator is willing to demonstrate (by compliance demonstration testing) that no increase in actual particulate matter emissions would occur.

"Actual" and "Potential" Increased Emissions - Determining whether NSR applies to a change is to determine whether it has resulted in a significant net emissions increase of any regulated pollutant under the Clean Air Act. If a modification is made, the comparison would be made between the present actual and future potential maximum. Therefore future potential emissions woild be calculated by running every day at 100% output must not be significantly greater than your old average emissions. The practical effect of employing this "actual-to-potential" methodology is that an emissions increase can be found from virtually any physical change, even in cases in which no actual increase in emissions has occurred.

Prior methodology for baseline determination was fairly restrictive. The two year period prior to the modification was defined as the designated baseline time frame. If this period was not reflective of normal operation (such as during a business slow down period), then another contiguous two years during the previous five could be selected.

---

[4]Federal Register, Volume 67, Number 251.

Under new regulations the operator can choose at his own discretion any contiguous 24 months in the last 10 years as the baseline period. Key items of this rule are:

- The methodology allows for an actual-to-future-actual emissions comparison.
- A margin of emissions related to demand growth of the original equipment may be factored out of the comparison.
- The baseline emissions are calculated according as the emissions of any contiguous 24 months in the past 10 years.
- The baseline cannot include emissions that may exceed present enforced guidelines, such as NO_x Reasonably Achievable Control Technology (RACT).
- Future emissions must be calculated and reported on a rolling 12 month basis for the next 5 years. (The implication here is that some sort of continuous monitoring will be required.)
- If the potential to emit increases (if the furnace is enlarged or otherwise capable of melting more glass), then the calculation period is extended to 10 years.
- If the operator chooses to use and can demonstrate compliance with the old actual to future potential analysis, no exceptional record keeping is required

"Debottlenecked" - EPA now wants to challenge industry's practices of "routine maintenance, repair and replacement" by looking at other processing equipment associated with production capabilities. EPA takes the position that the emissions increase implications from "debottlenecking" must be considered in determining NSR applicability. A facility engages in debottlenecking when it makes a physical or operational change that will allow units that are not changed to operate at a greater capacity than prior to the change. EPA may argue that changes in forming machines or bottling machines that increase the potential level of production from the line should be considered debottlenecking changes, that result in an increase in potential emissions from the furnace. The counter-argument from the facility's standpoint is that changes to bottling machines or forming machines are a part of normal operations and, therefore, are not a "physical change" or "change in method of operation" triggering NSR.

PSD Requirements - New and modified sources in attainment areas, *i.e.*, where air quality standards have been met (and in unclassifiable areas), are required to follow PSD rules. This includes obtaining a pre-construction permit, prove that the unit will not cause violations of certain air quality standards, and show that that its operations are in continuous compliance with the "Best Available Control Technology" (BACT) requirements.

In non attainment areas, *i.e.*, where air quality standards have not been met, new and modified sources are required to 1) obtain pre-construction permits, 2) offset emissions increases with emissions reductions from other sources in the area, and 3) install lowest achievable emissions rate technology ("LAER"). LAER is based on the most stringent emission limitation in any State's SIP, or achieved in practice by the source category under review.

There are three basic criteria in determining PSD applicability. The first and primary criterion is whether it is a "major" source or "major" modification. Source size is defined in terms of "potential to emit," and is major if it has the potential to emit any pollutant regulated under the Act in amounts equal to or exceeding specified major source thresholds which are predicated on the source's industrial category.

The second criterion for PSD applicability is that a new major source would locate, or the modified source is located, in a PSD area. A PSD area is one formally designated, pursuant to section 107 of the ACT and 40 CFR 81, by a State as "attainment" or "unclassifiable" for any criteria pollutant, i.e., an air pollutant for which a national ambient air quality standard exists.

The third criterion is that the pollutants emitted in, or increased by, "significant" amounts by the project are subject to PSD. A source's location can be attainment or unclassified for some pollutants and simultaneously non attainment for others. If the project would emit only pollutants for which the area has been designated NonAttainment, PSD would not apply.

MOST STRINGENT AGENCY REQUIREMENTS

In the past, the level of air pollution control was typically based upon the facility location and the status of the region relative to meeting Federal Ambient Air Quality requirements. The regions with the worst air, would adopt the most stringent standards. Typical of these regions have been in California - SCAQMD (Los Angeles), BAAQMD (San Francisco), and SJVUAPCD (San Joaquin Valley). San Joaquin in unique in that there are Container, Flat and Fiber glass facilities. Further, it is in non compliance for both ambient Ozone and fine particulate (PM $_{2.5}$). Many other regions in the U.S. have patterned their regulations to be consistent with this district's regulations.

SJVUAPCD Rule 4354 - The original rule was established in 1998 to limit glass furnace NO_X emissions. The glass container industry interacted with the District staff to formulate a rule which was achievable by process modification (4 lbs. NO_X per ton of glass melted). Compliance was required at the time of the next rebuild, which allowed furnace design changes. Because the region is still out of compliance with NAAQS, EPA Region IX has forced this District to significantly tighten all emission limits.

This rule is was modified Oct. 16, 2008 to include additional requirements, which are essentially BACT, and must be in place by 2011 or 2014 (depending upon the last furnace rebuild date). These limits are the most stringent being imposed upon existing furnaces in the U.S. The limits are based upon a block 24 hour block average or a 30 day rolling average:

	Container	Flat Glass	Fiber
Pounds per Ton of Glass Melted			
Total PM $_{10}$ Particulate[5]	0.50	0.50	0.70
SO_X	0.9 or 1.1[6]	0.9	1.7 or 1.2[7]
NO_X	1.5[8]	3.7 or 3.2[9]	1.3
CO	1.0 or 300 ppmv	0.9 or 300 ppmv	1.0 or 300 ppmv
VOC	0.25 or 20 ppmv	0.1 or 20 ppmv	0.25 or 20 ppmv

CAIR Background - On March 10, 2005 EPA promulgated the Clean Air Interstate Rule (CAIR). The CAIR requires certain upwind States to reduce emissions of nitrogen oxides (NOx) and/or sulfur dioxide (SO₂) that significantly contribute to non attainment of, or interfere with maintenance by, in

---

[5]Filterable and Condensable ("back half")
[6]Using >25 % post cousumer cullet and Oxy-fuel firing
[7]Rolling 30 day average
[8]Rolling 30 day average
[9]Rolling 30 day average

downwind States with respect to the fine particle and/or 8-hour ozone national ambient air quality standards (NAAQS). EPA is now requiring 28 States and the District of Columbia to revise their State implementation plans (SIPs) to include control measures to reduce emissions of SO_2 and/or NO_X. It is likely that, in the near future, these states may adopt the SJVUAPCD's Rule 4354 limits for glass furnace emission standards.

The first phase of NO_X reductions starts in 2009 (covering 2009-2014) and the first phase of SO_2 reductions starts in 2010 (covering 2010-2014). The second phase of both SO_2 and NOx reductions starts in 2015 (covering 2015 and thereafter). Each State covered by CAIR may independently determine which emission sources to control, and which control measures to adopt.

In March 2008, the U.S. Court of Appeals in Washington, DC. found the bulk of the rule "fundamentally flawed" and setting it aside. Many areas of the United States are currently in violation of the 1997 ozone and particulate matter standards. They are required to improve air quality, but if they cannot bank on the sulfur dioxide and nitrogen oxide reductions mandated by CAIR, these areas will not meet the standards on schedule.

The court upheld limited aspects of the rule, but it faulted CAIR on four major grounds:

1. CAIR's region-wide emission reduction and trading program is not designed to meet the statutory goal of ensuring that emissions from upwind states will not adversely affect air quality in downwind states.

2. CAIR does not provide sufficient protection to downwind states that were now in attainment with the particulate matter and ozone standards but could fall into non-attainment in the future.

3. This ruling effectively prevents EPA from setting lower emission limits for sulfur dioxide than those established by Congress in the 1990 act.

4. The 2015 deadline for CAIR's second phase of emission reductions was not in keeping with the act because it would come too late to help many downwind states meet 2010 attainment deadlines for the particulate matter and ozone national ambient air quality standards.

The court decided to vacate CAIR in its entirety because the rule was unlikely to survive "in anything approaching recognizable form," and "EPA must redo its analysis from the ground up." By vacating the rule, the court relieved utilities from any obligation to reduce sulfur dioxide and nitrogen oxide emissions.

In April 2005 EPA determined that 36 areas, with a total population of nearly 90 million people, contain above-standard $PM_{2.5}$. By April 2008, these areas were required to submit state implementation plans demonstrating how they would meet the standard, which must be achieved by April 2010 unless EPA grants an extension, which can be no more than five years. For Ozone, EPA designated 57 areas, with a total population of nearly 132 million people, as being in non-attainment with the eight-hour standard. State implementation plans for these areas were due to be submitted to EPA by June 2007. The deadlines for implementing the standard vary depending on the severity of non-attainment, but the great majority of areas are required to comply with the standard between 2009 and 2013.

OTC Background - The (OTC) is a multi-state organization created under the Clean Air Act (CAA). Their responsibility includes advising EPA on transport issues and for developing and implementing regional solutions to the ground-level ozone problem in 12 Northeast and Mid-Atlantic states.

Because ambient air quality in this region is often influenced by emissions from "upwind" states, the Commission developed Model Rules to be implemented for the entire Region. Their Technical Committee has compiled a list of Reasonably Available Control Technology (RACT) Control Technique Guidelines (CGT's) categories. The OTC Model Rule for NO_X have been recommended as the starting point for RACT updates for State adoption in future rule making. Pennsylvania is adopting the OTC's specific recommendations.

RECENT EPA "NEW SOURCE" RULE

Effective July 15, 2008[10], EPA issued final rules amending the New Source Review ("NSR") permitting program for $PM_{2.5}$. This rule completes a framework for the review and issuance of preconstruction permits to meet the National Ambient Air Quality Standard ("NAAQS") for $PM_{2.5}$. *It also* addresses the use of PM_{10} as a surrogate for $PM_{2.5}$ and established key elements to implement the $PM_{2.5}$ NSR program:

Major and minor NSR permits must address direct $PM_{2.5}$ emissions as well as $PM_{2.5}$ precursor pollutants.

- Sulfur dioxide ("SO_X") must be regulated as a $PM_{2.5}$ precursor.
- NO_X must be regulated as a $PM_{2.5}$ precursor, unless a State demonstrates NO_X is not a significant contributor to the formation of $PM_{2.5}$ for an area.
- Volatile organic compounds ("VOC") and ammonia are not regulated as $PM_{2.5}$ precursors unless a State demonstrates VOC or ammonia is a significant contributor to $PM_{2.5}$ formation in an area.

A new source located in an attainment area is considered a major source for Prevention of Significant Deterioration ("PSD") permit program purposes if it is one of 28 listed source categories with potential emissions of 100 tons or more per year of a regulated NSR pollutant, including a $PM_{2.5}$ precursor, or 250 tons or more per year of a regulated NSR pollutant, including a $PM_{2.5}$ precursor, if it is a non-listed source. A new source in a nonattainment area is a major source if its potential emissions of a regulated NSR pollutant, including a $PM_{2.5}$ precursor, are 100 tons or more per year.

An existing source making modifications is subject to NSR if the modification results in a significant emissions rate, defined as:

- Direct $PM_{2.5}$– 10 tpy
- SO_2 precursor – 40 tpy
- VOC precursor – 40 tpy (if regulated)
- Ammonia precursor – as set by State

Preliminary estimates based on EPA's most recent data indicate that approximately 250 facilities apply for a PSD or non attainment NSR permits annually. There are approximately 20,000 sources that would be classified as major under the Clean Air Act, and most of the 375 glass melting furnaces in the U.S. are considered "major" sources. EPA has become increasingly concerned about how few glass furnaces have installed state-of-the-art control technology.

EPA headquarters coordinates discussions with its regional offices regarding NSR enforcement against the glass industry. In a unified approach, the EPA regional offices are specifically targeting certain

---

[10]73 Federal Register 28321 on May 16, 2008

industries, including the Glass industry. Their investigations begin with process equipment changes since implementation of their NPSP regulations went into effect (1980 for Glass). In the last 3 years, a number of glass companies have received Clean Air Act "section 114" information requests relating to NSR and NSPS compliance. More recently, glass facilities have been issued Notice of Violations (NOV) based upon EPA's interpretation of modifications they believed should have triggered PSD/NSR.

Many states are now requesting NSR permitting advice from EPA as a result of increased pressure from EPA. Some states are taking a very narrow view of what constitutes "routine maintenance, repair and replacement" under the NSR regulations, and that EPA has been asking questions in its section 114 requests about "efficiency" improvements and "debottlenecking" activities. Some states are requesting that facilities install continuous monitoring devices as a condition to obtaining minor pre construction permits, even if the facilities are not required to install control technology under the major NSR program

Largely because of EPA's influence over regional or state authorities, a number of glass associations are cooperating in sharing their experiences with agency rule interpretations. Because of the complexity of understanding permit requirements and serious consequences of being considered out of compliance, most glass manufacturers are turning to experts in environmental law for guidance on permitting issues.

AB-32 BILL

The Global Warming Solutions Act of 2006 requires that California reduce its greenhouse gas emissions to 1990 levels by 2020. The 1990 Greenhouse Gas Emissions Level establishes the actual number of tons of emissions that California is required to reach.

California's AB 32 bill requires a 25 percent cut in GHG to be reduced within the state's borders by 2020 in order to bring the total down to 1990 levels. In at least eight other states, political momentum is building to take similar steps to limit emissions of greenhouse gases linked to climate change, a trend that could increase the pressure for a national system.

The glass container manufacturing's GHG emissions consists essentially only of CO_2. The two sources are from Natural Gas combustion (entire facility) and the fusion loss in melting of Carbonate raw materials (Limestone and Soda Ash). Furnace efficiency and cullet levels will vary emissions, but generally for producing 100 tons of glass - Natural gas contributes ~ 30 tons of CO_2, and ~ 10 tons of CO_2 from raw materials.

The mandatory reporting regulations require annual reporting from facilities in the state that emit over 25,000 tons of carbon dioxide each year from on-site stationary combustion sources . This includes the typical glass manufacturing facility. Affected facilities will begin tracking their emissions in 2008, to be reported beginning in 2009. Verification will take place annually or every three years, depending on the type of facility. ARB is also developing a training and accreditation plan for third-party verifiers.

On Oct. 22, 2008 CARB indicated their interest in promoting a "Cap & Trade" program for the glass industry in California. At this time, the details have not been formalized, but the program is targeted to go into effect by 2012. Similar perspectives to use a Cap & Trade system on the Federal level are also expected.

GLOBAL WARMING AND SUSTAINABILITY OF THE GLASS INDUSTRY

Guy Tackels
Saint-Gobain Conceptions Verrières

ABSTRACT

In 2008, the European glass industry has to face high energy costs and high carbon constraints in a world imposing tremendous reductions of GHG[1] emissions. In the mean time, the glass industry has also to adapt the production to ever tighter environmental constraints like the so-called European IPPC directive and REACH regulations. However, the climate change issue is also a chance for the glass industry because a growing market can be foreseen for more and more sophisticated glass products, which will allow energy saving and reduced CO_2 emissions.

The glass industry belongs to the category of "energy intensive industries", because energy represents a large part of the manufacturing cost, typically 10 to more than 20% for glass products. For a long period, energy consumption was clearly connected to the GDP and fossil energy was considered as abundant. Wood, coal, oil and gas were successively used with no other particular attention than to reduce the cost of energy as it was an important part of the total manufacturing cost. The first alert was in the seventies, with the first and second oil shocks. Oil became scarce and expensive and the idea that this natural resource could disappear came for the first time to the front of the news.
In the 1990's and up to now, the environment has become a major preoccupation and since 1997, with the signature of the Kyoto Protocol, climate change is also a top priority.
Today, we witness a fantastic growth of industrial production in the non OECD countries (namely China and India) and burning fossil energy is considered as a threat for the climate. It is suddenly realised that fossil energy will become again scarce and expensive for a long time. And in many countries, burning fossil fuels becomes more difficult, because it is necessary to respect more and more severe legislations on air quality, water quality, waste and chemicals.

In this article, we try to answer to the following questions.
- What are the options for a glassmaker taking into account that 80% to 90% of the energy is used today in the form of fossil energy and the rest as electricity?
- Which evolution is required for further energy savings, CO_2 reduction and pollution control in the manufacture of glass?
- Which opportunities for glass products in a world with high carbon constraint?

ENERGY SAVINGS AND CO_2 REDUCTION

Since 1960 the average specific energy consumption of the European Glass Industry has been reduced roughly by 50% to 60%.
The main driving force for this progress can be found in the fact that the reduction of energy consumption was always and continues to be a priority for the glassmaker, because energy costs represent a substantial amount of the production cost. Technically, the progress was due to a continuous increase of glass recycling starting after the first oil shock and improvements of the melting technology: better furnace design allowing increased specific pull, better energy recovery and less

---

[1] Acronyms are detailed in a glossary at the end of the paper

energy losses. The average size of the installations was also increased (causing decreasing specific energy consumptions) in order to respond to the increased capacities required by the market.

The good performance in CO_2 reduction is also explained by the continuous tendency to switch from heavy fuel oil to natural gas. This trend to use more gas was mainly due to more stringent environmental regulation (namely on SO_2) since 1995 in several European countries (France, Germany, The Netherlands).

WHICH ENERGY FOR THE FUTURE?

One of the key factors concerning the environment is now the climate change. Today, almost everybody agrees on the reality of climate change. There is a global consensus that climate change is a significant threat to the world and that the changes are mostly due to human activities. There is also evidence that climate changes are occurring and that costs associated with damage are considerable and precaution and action are urgently needed. It is the reason why some EU member States like Germany, UK. The Netherlands and France are politically committed to very ambitious reduction targets, up to 75% of GHG reduction until 2050 compared to 1990. This reduction will be required in developed countries in order to maintain the average world temperature below a 2°C increase.

It is well recognised that tremendous efforts have to be made in all sectors of the society and not only in the industry which already achieved, like glass manufacture, much progress. In particular, households and transports have to reduce efficiently their own energy consumption and related CO_2 emissions.

In this context, we can wonder what will be the situation of the glass industry in 2050.

The answer is of course very difficult. What will be the cost and availability of fossil energy and which one will be available at a reasonable cost in 2050? Will the glass industry use biomass or come back to (green) coal? In the last case, will the CCS (Carbon Capture and Storage) be a technique valid for relatively small installations as a glass furnace or a glass plant?

We believe in the statement made by IEA (International Energy Agency), that fossil energy will continue to be the first source of primary energy for a long period and glass will use this energy at least for 20 or 30 years. And after?

For glass, a technical solution is already available and used in some particular cases: the all electric furnace. However, several conditions must be fulfilled if we want to see the biggest part of the glass industry i.e. soda-lime glass for flat glass and containers switching to all electric melting:

- Electricity must be produced carbon free (nuclear or renewable like hydraulic, biomass, etc...) and competitive with the fossil energy, including the carbon constraint. It means probably that carbon price will be very high and electric energy from fossil fuels (without CCS) becomes unaffordable. Shall we see one day photovoltaic panels on the roof of our plants, and wind mills around the installations of glass production?
- The carbon constraint must be equally distributed around the world. It means a worldwide post-Kyoto agreement on the subject after 2012. The Kyoto Protocol is only a very modest first step. A consensus on the future of the Kyoto Protocol will however require time and hard political agreement.
- This solution must be globally competitive with other materials (for instance in the packaging sector).

Even in absence of an international post-Kyoto agreement, the carbon constraint is thus installed in Europe for a long time, not to say for ever.

REVISION OF EU ETS DIRECTIVE: A DIFFICULT CHALLENGE FOR 2020

Let us now focus on the relatively short term i.e. the year 2020.
A tool has been developed in Europe in order to help the industries to fulfil their obligation to reduce their GHG emissions with a minimum overall cost. The first period of functioning (2005-2007), sometimes considered as "learning by doing period" showed contrasted results.
As preliminary remarks, we can observe:

- The EU ETS is largely dominated by the energy sector (above 50% of the market). The EU ETS system is not tailored for the glass industry which represents annually only 21.6 Million metric tons on a total of 2.1 billion tonnes of CO_2, it means 1% of the market. The specificities of the glass production, which are very diverse, are not taken into account.
- The cap and trade system for CO_2 emissions set up by the Commission and agreed at EU Parliament level results in fixing a cap on production. Absolute targets for CO_2 are fixing limits to the production itself and do not take into account the requirement to adapt production to the market demand.
- The NAPs (National Allocation Plans) were not ambitious enough, because they had to include optimistic production forecasts.
- As a consequence, and because the EU governments wanted to maintain the competitiveness of their industries, most EU countries were "long" (i.e. they had an excess of allowances for the period 2005-2007) and in May 2006, when the reality became public, the price of CO_2 collapsed to nearly 0 €/t.

To achieve the global warming objective, in January 2008 the European Commission proposed an "energy-climate" package of legislation, now based on four pillars: the fair division among Member States of efforts to reduce greenhouse gases, the promotion of renewable energy, the capture and storage of CO_2 and the improvement of the European Trading Scheme (EU ETS) including harmonisation of rules, monitoring and allocation. In this package, the EU ETS is the central piece of European Commission's proposal and several modifications are foreseen to improve the efficiency of the current system. The main principles in the revised text are the following:

- A cap & trade system is maintained
- Duration of the period is increased to 8 years
- Allowances will be granted per sector (e.g. glass industry) at European level
- Full auctioning will be the rule unless you can prove that your sector will be submitted to severe non-EU competition (carbon leakage) in case of purchasing all requested CO_2 allowances;
- Benchmarking will be used when it is possible to set the free allocation for sectors with world-wide competition (outside EU) exposed to carbon leakage aspects.

Today, the allocation methods of allowances are proposed by Member States and the scheme is validated by the Commission. Beginning in 2013, the Commission will assume the full responsibility, thereby de facto limiting the differences in the treatment of industrial installations between countries. If a satisfactory international agreement (main emitting countries worldwide) is signed, the minimum European objective, of a 20% reduction of GHG emissions by 2020 from 1990 level, would increase to a 30% reduction. The effort would have to be made primarily by the sectors covered by the EU ETS where the cost of reducing emissions is considered to be lower. As a consequence, the absolute cap for EU ETS sector will be -21% from 2005 emission level.

A second major modification relates to the allocation principle. Allowances would no longer be free but would be auctioned. Electricity sector (above 50% of the EU ETS emissions) will have to purchase all of their allowances beginning in 2013. For other sectors (mainly manufacturing industry) the Commission is proposing to gradually increase the percentage of allocation sold at auction from 20 % of the cap in 2013 to 100% in 2020. It remains to be determined how this huge amount of money (around 60 to 80 billion € annually) would be distributed among the States and recycled for climate change mitigation and to help industry to reduce their CO_2 emissions in an efficient way.

The absolute emissions reduction targets set for 2020 are extremely ambitious. The steady reduction in allocations and the auctioning of an increasing percentage of allowances will induce a cost increase for CO_2 (due to increasing production and lack of cost-effective energy saving technologies) resulting in higher production costs. This increase of production costs could be problematic for the competitiveness of some sectors. The system is designed to promote investments in technologies that produce fewer emissions. The problem comes from the distortions to competition with installations outside the EU, which are not subject to the carbon constraint. The result could be the relocation of certain production operations (cement, steel, glass,...) in a country not submitted to carbon reduction. This artificial reduction of European emissions is called "carbon leakage" and will seriously limit the effectiveness of the market. Consequently the environmental target will be partially or even completely missed or even counter-acted.

One of the solutions could be to give a certain percentage of free allocation (based on benchmarks) to the sectors affected by the international competition. However, it is not so easy to establish the right criteria and the methodology to assess which sectors or sub-sectors and activities would be exposed to significant risk of carbon leakage.

THE CONSEQUENCES FOR THE GLASS INDUSTRY

The carbon constraint will be present for a long time (or forever) and we need, as glass industry, to participate to the efforts required for climate change mitigation. However, the best way to reach this target has still to be found.

First of all the glass industry growth must be taken into account. According to the historical data and taking into account the growth of some sub-sectors boosted by the necessity to reduce CO_2 emissions (in household sector namely), we can easily foresee an average economic growth above 1% per year. Taking into account the cap (-21% compared to 2005) the total reduction needed in 2020 and expressed in specific value (tonne of CO_2/tonne of glass) will be above 40%.
As an example, let us take an average value for the specific energy consumption of an end fired furnace in the glass container sector: 4.2 GJ/tonne (3.5 M BTU/ton). A reduction of 40% means to obtain a specific energy consumption of 2.52 GJ/tonne (2.2 M BTU/ton). This value is not so far from the theoretical energy requirements for melting glass. It is of course impossible to obtain technically this level of energy consumption by conventional means. We can conclude that EU ETS obligations will be difficult to fulfil. The glassmaker will have the choice between the following alternatives:
- To reduce the production accordingly. This is not acceptable.
- To produce glass in a country where the CO_2 constraint does not exist. It means carbon leakage.
- To buy CO_2 allowances. The CO_2 price will be probably very high and pass through not always possible (competition from other materials not exposed to EU-ETS or competition from outside Europe).

- To reduce CO_2 emissions by technical means. For instance we can choose to increase substantially electric boosting. This solution is probably not acceptable because electricity cost will also rise sharply due to the carbon constraint unless electricity will be produced by other source than fossil fuels.. We will need a technology breakthrough. One possibility will be to use carbon free energy including biomass or other kind of renewable energy. This will require long term R&D before to be really available.

We believe that the long term solution (after 2020-2030) remains the "carbon free" electricity. However this solution will be viable only if electricity price is not anymore connected to CO_2. For instance if CCS becomes a valid operational solution, electric melting could become an attractive solution for the glass industry. However in the mean time other solutions linked to renewable energy must be found urgently.

Unless there is a technological or economic breakthrough (like the generalisation of electric furnaces mentioned above), we do not see how the glass industry will be able to respect the required reduction in 2020. The glass industry will then be obliged to buy huge amounts of allowances. In fact, over the last 20 years, the progress made on specific emissions was just able to compensate on average the increase of production volume.

As a first conclusion, we can say that the target fixed by the Commission is not realistic.

In the scope of an international post-Kyoto agreement an alternative solution is proposed by some industrial sectors namely cement and steel. "Sectoral agreements" have been quoted as a way forward in the fight against Climate Change. A "sectoral agreement" is an agreement that companies representing a significant market share of a homogeneous sector in a given perimeter, will achieve within this perimeter and over a certain period of time, a precise and verifiable progress in the fight against Climate Change. Multinational companies already face diverse national climate change schemes e.g. EU ETS in Europe and the American RGGI. It can be anticipated that the diversity of the systems will increase and a "sectoral agreement" could be a useful tool for industry.

However some structure like an international federation or an international institute must be available worldwide as it is the case for cement and steel, but not for glass. It is maybe time to build this structure for the glass industry.

As mentioned earlier carbon price will probably increase sharply in the future. Some techniques will become more attractive like oxy-firing for several types of glass, batch/cullet preheating in container sector, etc. However we do not see, on average, rapid significant modifications of the present trend and changes in making economical decisions on energy saving investments (pay back times are considered to be too long). Today, we do not receive the right signal leading to the required investment for a substantial emission reduction. This is the more perverse effect of the system: penalisation of the companies with substantial growth and (small) financing of declining markets. Nobody is really a winner with such results and the target on climate change improvements is completely missed. A big mistake is to mix completely different sectors with different economic models, competition conditions and technical emission reduction possibilities. Up to now the system failed also because of the situation observed in the electricity market. Manufacturing industry is penalised a second time by the tremendous increase of electricity price observed on the liberalised electricity market due to the inclusion of the CO_2 cost into the price of electricity.

We see however a positive point with EU ETS, but this is not connected to the cap and trade system and not at all to full auctioning. The carbon constraint which is inevitable pushes the industry including glass to boost R&D on CO_2 emission reduction and new ages of glass melting can be predicted.

Finally some allowances will be distributed for free in the period 2013-2020, based on a benchmarking. A benchmark is rather difficult to implement in the glass industry because of the diversity of situations and close negotiations with the authorities would be anticipated. Many perverse effects have to be avoided. For instance, the choice of the melting technique can not be guided only by the CO_2 constraint. We must keep some flexibility for the type of furnace, the type of energy we use (oil or gas) because these choices, guided by economic reasons, could be critical.

Many ideas exist to improve the existing system. However a long time will be necessary to implement worldwide a system adapted to the specificities of our industry. The representatives of the glass industry and the European Commission have to work together in order to make positive proposals for improving the EU ETS. We need a well performing tool and not a cumbersome system which misses the initial target to smooth the carbon constraint and to maintain competitiveness.

GLASS PRODUCTS AND CLIMATE CHANGE

The other side of the coin is also worth investigating. What saving can be obtained thanks to the use of glass products?

The glass industry can provide numerous answers to the environmental challenges of climate change. These products, which are more and more sophisticated, offer solutions for limiting GHG emissions and, in the field of renewable energy, providing materials that make the most efficient use of natural resources. The glass industry supplies building materials (flat glass, glass wool insulation) that, once installed, allow far more energy to be saved than was required for their manufacture. We can refer to different studies made at European level. They therefore contribute for both energy saving and reduction of CO_2 emissions.

In conclusion, the CO_2 emitted by the manufacture of glass for building materials is greatly outweighed by the potential saving.

The EU Union has also an ambitious target for renewable energy production and it is a political priority. In this field, glass has also an important role to play. Glass continuous filament industry provides threads and fabrics for reinforcing wind turbine blades. The sector is growing fast and cost efficiency is improving continuously.
In the field of solar energy, a distinction must be made between thermal and photovoltaic applications. In both case, glass plays a fundamental role. The challenge today is to decrease the cost of production.
In all these applications in renewable energy, the energy used to manufacture glass is very small (energy pay back times of far less than 1 year) compared to the saving of primary fossil fuels by using renewable energy.

CONCLUSION

In the future, the glass industry will have numerous opportunities for market developments. On the production side, reduction of CO_2 emission remains a priority and modifications of EU ETS are required if we want to have an efficient tool at our disposal. Economic solutions to reduce significantly the CO_2 emitted at the furnace level are still to be found. "Carbon free" electricity and renewable (biomass namely) will probably replace fossil fuels at least partially.

GLOSSARY:

CCS: Carbon Capture and Storage
EU ETS: EU Emission Trading Scheme
GDP: Gross Domestic Product
GHG: Green House Gas
IEA: International Energy Agency
IPCC: International Panel on Climate Change
IPPC: Directive on Integrated Pollution Prevention and Control
NAP: National Allocation Plan
OECD: Organisation for Economic Co-operation and Development
REACH: Registration, Evaluation, Authorisation and Restriction on Chemicals
RGGI: Regional Greenhouse Gas Initiative